Deprenyl – Past and Future

Edited by
W. Kuhn, P. Kraus, and H. Przuntek

Journal of Neural Transmission
Supplement 48

SpringerWienNewYork

Prof. Dr. W. Kuhn
Dr. P. Kraus
Prof. Dr. H. Przuntek
Neurological Clinic, St. Josef Hospital, Bochum,
Federal Republic of Germany

© 1996 Springer-Verlag/Wien

Typesetting: Best-set Typesetter Ltd, Hong Kong
Printing: A. Holzhausens Nfg., A-1070 Wien
Graphic design: Ecke Bonk
Printed on acid-free and chlorine-free bleached paper

With 16 Figures

ISBN-13: 978-3-211-82891-5   e-ISBN-13: 978-3-7091-7494-4
DOI: 10.1007/978-3-7091-7494-4

Foreword

L-Deprenyl, N,α-Dimethyl-N-2-propynyl-, (R)-(9CI) also called (−)-Deprenyl, R-Deprenyl or E-250, is now designated "Selegiline" (INN).

Selegiline potently, dose-dependently and highly selectively inhibits monoamine oxidase type B (MAO-B), without inducing the cheese effect.

Although it was developed as an antidepressant/psychostimulant in the late sixties and early seventies, the first trials in patients with Parkinson's disease (PD) (Birkmayer et al., 1975) led to a completely new view of its potential. It became a "must" in the therapeutic handling of PD, when it was shown that Selegiline was able to protect dopaminergic neurons from the toxicity of N-methyl-4-phenyl-1,2,3,6-tetrahydropyridine (MPTP). In the following years, Selegiline was marketed as a neuroprotective substance, a fact supported by experimental studies, but later questioned when long-term clinical studies did not prove unequivocally this efficacy in humans with PD.

Although not an ideal drug with regard to its structure and metabolism Selegiline has gained a respectable scientific reputation during the last 20–25 years.

The varied clinical designs employed in long-term trials in PD, as well as recent experimental data indicating that Selegiline facilitates expression of various trophic factors underline the fact that with regard to its clinical pharmacology, this drug is one of the most thoroughly examined CNS compounds used in PD. It is therefore clear that Selegiline is the "gold standard" in evaluating drugs for the treatment of PD in clinical long-term trials, as well as in evaluating drugs with neuroprotective/neuroregenerative potential.

From time to time it is necessary to examine the "status quo" of a drug, and particularly of one that exhibits so many interesting and important characteristics in both the clinic and the laboratory. Specialists from around the world therefore discussed the history and the future of Selegiline. Interesting lectures and highly stimulating discussions illuminated Selegiline's role in the context of other drugs used in PD, and in particular its relation to emerging new therapeutic strategies.

This small conference which took place at Lake Starnberg, Bavaria, in spring 1995, touched on some important and current issues; e.g., the question of the toxicity of its metabolites (comparing Selegiline's properties and those of Rasagiline, a compound not metabolized to amphetamine and methamphetamine), the role of Selegiline in the inhibition of apoptosis, and its role in the induction of the expression of neurotrophic factors.

In contrast to other drugs used to treat PD, investigation of Selegiline has led to a variety of new therapeutic concepts and insights into the pathological mechanisms of neurodegenerative disorders. While these proceedings are prepared, it is evident that research with regard to neuroprotection and neuroregeneration continues to advance and that Selegiline remains the major substance in this field, with which novel substances must be compared.

Würzburg, autumn 1996   **P. Riederer**

Preface

Over the past 30 years considerable information regarding the central role of monoamine oxidase (MAO) in regulating monoaminergic activity has accumulated. The discovery that L-Deprenyl is a selective irreversible inhibitor of MAO-B (Knoll and Magyar, 1972; Youdim et al., 1972) has initiated the first clinical study in Parkinson's disease (Birkmayer et al., 1975). In this trial a therapeutic efficiency of L-Deprenyl as an adjunctive agent to levodopa has been described. This was confirmed in various clinical investigations in later years. The clinical effect of L-Deprenyl was explained on the basis of MAO-B inhibition and subsequent enhancement of dopaminergic neurotransmission.

In recent years new experimental data have challenged this concept. In vitro and in vivo studies are suggesting that L-Deprenyl may have neuroprotective and/or neuroregenerative properties, too. Further, two long-term clinical studies (DATATOP, SELEDO) have brought forward a new discussion on the possible mode of action of L-Deprenyl in Parkinson's disease and various other neurologic and psychiatric disorders. In the light of these new and fascinating experimental and clinical results this symposium was initiated to provide a forum for intensive discussion on new horizons in basic biochemistry, pharmacology and clinical research of Parkinson's disease.

Bochum, October 1996

W. Kuhn
P. Kraus
H. Przuntek

Contents

Listed in Current Contents/Life Sciences

J Neural Transm (1996) [Suppl] 48: 1–6

New aspects of pathology in Parkinson's disease with concomitant incipient Alzheimer's disease*

H. Braak[1], E. Braak[1], D. Yilmazer[1], R. A. I. de Vos[2], E. N. H. Jansen[2], and J. Bohl[3]

[1] Zentrum der Morphologie, J.W. Goethe Universität, Frankfurt/Main, Federal
Republic of Germany
[2] Streeklaboratoria voor pathologie, Burg. Edo bergsmalaan, Enschede,
The Netherlands
[3] Abteilung für Neuropathologie, J. Gutenberg Universität, Mainz,
Federal Republic of Germany

Summary. Alzheimer's disease and Parkinson's disease are the most common age-related degenerative disorders of the human brain. Both diseases involve multiple neuronal systems and are the consequences of cytoskeletal abnormalities which gradually develop in only a small number of neuronal types. In Alzheimer's disease, susceptible neurons produce neurofibrillary tangles and neuropil threads, while in Parkinson's disease, they develop Lewy bodies and Lewy neurites. The specific lesional pattern of both illnesses accrues slowly over time. Presently available data support the view that fully developed Parkinson's disease with concurring incipient Alzheimer's disease is likely to cause impaired cognition.

Parkinson's disease (PD) is a common neurodegenerative disorder in which specific subsets of dopaminergic melanin-laden projection neurons of the substantia nigra, as well as many other neuronal types, undergo premature cell death (Braak and Braak, 1986; Gibb, 1991). A further frequently occurring multisystem disorder of the ageing human brain is Alzheimer's disease (AD). The two diseases differ not only in their clinical symptoms, but also with regard to the pathologic brain changes. Common to both diseases is that they are disorders of the cytoskeleton of a few susceptible neuronal types. In AD, the cytoskeletal changes are composed of neurofibrillary tangles (NFTs) and neuropil threads (NTs) (Braak et al., 1994a), while in PD, they consist of Lewy bodies (LBs) and Lewy neurites (LNs) (Braak et al., 1994b). Nerve cells developing the cytoskeletal alterations die, for reasons yet undiscovered (Braak and Braak, 1991, 1994; Braak et al., 1994b, 1995, 1996; Fearnley and Lees, 1994; Jellinger, 1990, 1991, 1994; Lowe, 1994).

---

*The results are described in detail elsewhere (Braak et al., 1996)

A specific and bilateral lesional pattern is developed during the course of both disorders. Initial changes occur consistently at specific predilection sites. As the severity of destruction increases, additional nuclei and areas gradually become involved, Many neuronal types, cortical areas and layers, as well as subcortical nuclei remain unscathed, while others exhibit severe destruction. The PD- and AD-specific lesional pattern is remarkably consistent across cases with only minor interindividual variation (Braak and Braak, 1991; Braak et al., 1993, 1994b, 1995, 1996; Jellinger, 1991). The pathologic process underlying AD preferentially destroys "afferent" cortical structures such as the entorhinal region and the neocortical association areas. In contrast, "efferent" subcortical structures such as the central nucleus of the amygdala and the substantia nigra are the main targets of destruction in PD. A substantial number of the affected nuclei and areas are components of the limbic system.

In AD, six developmental stages can be distinguished based on the predictable sequence of spreading of the neurofibrillary changes throughout the cerebral cortex (Braak and Braak, 1991). Initially, a few projection cells of the transentorhinal region are affected (clinically silent transentorhinal stages I and II). The pathologic process then proceeds to other cortical and subcortical components of the limbic system (limbic stages III and IV), and eventually extends into the neocortex (neocortical stages V and VI). Stages III and IV represent incipient AD, while stages V and VI correspond to fully developed AD (Bancher et al., 1993; Braak et al., 1993; Braak and Braak, 1994; Jellinger et al., 1991).

In the normal human brain, somato-sensory, visual, and auditory data proceed through core and belt regions of the parietal, occipital, and temporal neocortex to the prefrontal association areas. The major routes back from this highest organisational level of the brain to the premotor cortex (frontal belt) and the primary motor field (frontal core) are the striatal and cerebellar loops (Fig. 1a) (Alheid et al., 1990; Braak and Braak, 1993). The limbic loop system is a parallel circuit joining sensory association areas and prefrontal associations areas. Part of the sensory information branches off to converge upon the allocortex and amygdala (Felleman and van Essen, 1991). There, the data is processed and eventually conveyed to the prefrontal association cortex (Fig. 1b). The transentorhinal region and the lateral nucleus of the amygdala are the major ports of entry of neocortical data into the limbic system (Amaral et al., 1992). Projections from the hippocampal formation, the entorhinal region, and the basal and accessory basal nuclei of the amygdala contribute to the efferent leg of the limbic loop heading towards the prefrontal cortex. Some of these projections terminate in the ventral striatum, the "limbic" subdivisions of putamen and the accumbens nucleus (Alheid et al., 1990; Heimer et al., 1982). This input is supplemented by projections originating from the midline nuclei of the thalamus. The limbic data is then conveyed to the prefrontal cortex via the ventral pallidum and mediodorsal thalamic nucleus. In addition, the amygdala is reciprocally connected to the magnocellular portion of the mediodorsal thalamic nucleus and the mediofrontal and orbitofrontal areas. The agranular insular cortex is interposed in the efferent leg of the limbic

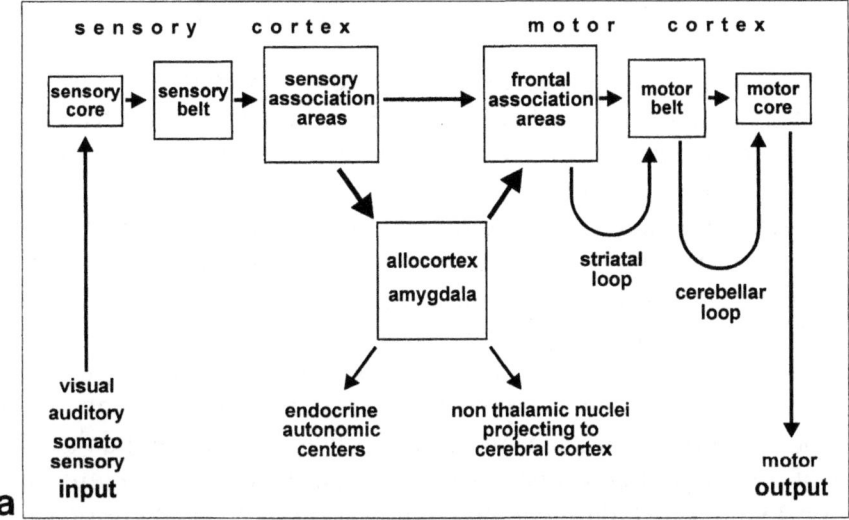

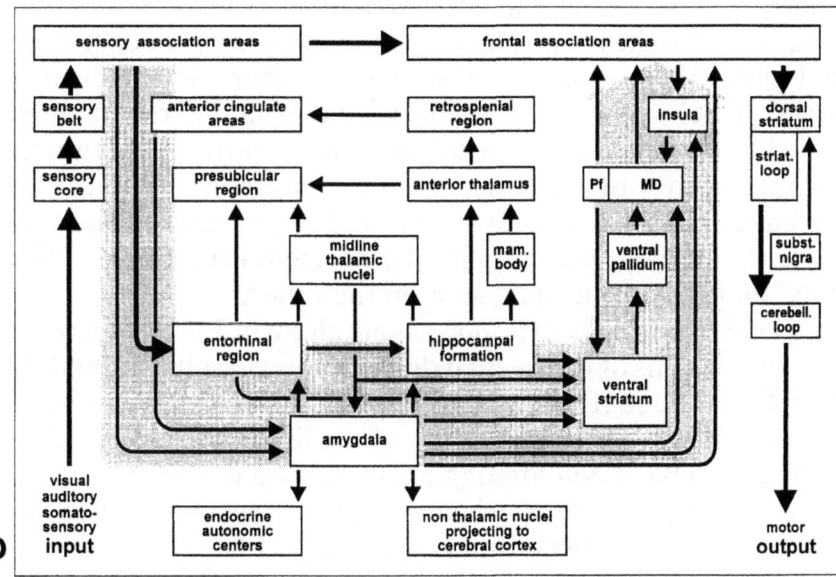

Fig.1. a The parietal, occipital, and temporal neocortex is each comprised of a core field, a belt region, and extensive association areas. Somato-sensory, visual, and auditory information proceeds through core and belt regions to a variety of association areas and then the data is transported via long cortico-cortical projections to the prefrontal association fields. Tracts generated from this highest organisational level of the brain guide the data back via the frontal belt areas to the frontal core field, the primary motor area. The striatal and cerebellar loops provide the major routes for this transport from the prefrontal cortex to the primary motor field. **b** Part of the stream of data from the sensory association areas to the prefrontal cortex branches off and eventually converges upon the entorhinal region and the amygdala (afferent leg of the limbic loop). The transentorhinal region and the lateral nucleus of the amygdala serve as major gates of entrance for this highly processed neocortical information. Via connections with the hippocampal formation, the neocortical data is then distributed to a variety of related limbic structures. Projections from the hippocampal formation, the entorhinal region, and the amygdala contribute to the efferent leg of the limbic lop which is directed to and exerts important influence upon the prefrontal cortex. The amygdala integrates exteroceptive sensory data with interoceptive stimuli from autonomic centers. A large number of amygdalar efferents terminate in nuclei regulating endocrine and autonomic functions. In addition, the amygdala generates efferent connections to all non-thalamic nuclei which in a non-specific manner project upon the cerebral cortex. *cerebell. loop* cerebellar loop; *mam. body* mamillary body; *MD* mediodorsal nucleus of thalamus; *PF* parafascicular nucleus; *striat. loop* striatal loop; *subst. nigra* substantia nigra

loop, as well. It receives projections from ventromedial portions of the basal nucleus of the amygdala and from the prefrontal cortex, and transfers data to the mediodorsal thalamic nucleus which, in turn projects to the prefrontal cortex (Amaral et al., 1992).

A characteristic pattern of extranigral lesions develops in PD. Important components of the limbic system are involved, including the entorhinal region (deep layer pri-α), the hippocampal formation (CA2-sector), the midline nuclei of the thalamus (in particular: reuniens nucleus and thalamic paraventricular nucleus), the anterior cingulate areas and agranular insular cortex (layer VI), the accessory cortical nucleus, and the ventromedial divisions of both the basal and accessory basal nuclei of the amygdala. The central nucleus of the amygdala is severely involved, as well. This nucleus has a strong impact on 1) all non-thalamic nuclei which project diffusely upon the cerebral cortex, and 2) many nuclei which regulate endocrine and autonomous functions (Fig. 1b). All these amygdala-dependent structures exhibit severe PD-related lesions themselves, as well. Hence, there is involvement of the cholinergic nuclei of the basal forebrain, the histaminergic tuberomamillary nucleus, the dopaminergic nuclei of the ventral tegmentum, the serotoninergic anterior raphe nuclei, and the noradrenergic locus coeruleus on the one hand and similarly severe destruction of the nucleus of the stria terminalis, the periaqueductal gray, the peripeduncular nuclei, the reticular formation, and the dorsal vagal area on the other.

The specific pattern fo extranigral pathology in PD does not usually produce any ouvert signs of intellectual deterioration. Similarly, mild AD-related pathology (up to stage III) may be asymptomatic as well. Superimposed upon the fully developed PD-related lesions of the limbic system, a mild additional AD-lesion may have a potentiating effect, adding considerably to the disturbances in the limbic loop, which may lead to the appearance of cognitive decline (Bancher et al., 1993; Braak and Braak, 1990; Jellinger and Bancher, 1995; Jellinger et al., 1991). This does not rule out that, in individual cases, other causes may be responsible for dementia in PD, i.e. co-occurrence of argyrophilic grains (Braak and Braak, 1989), of multiple infarctions, or of other lesions. Presently available data support the view that co-occurrence of AD-related brain destruction is the most common cause of intellectual decline in PD.

Acknowledgements

This study was kindly supported by the Deutsche Forschungsgemeinschaft, the Bundesministerium für Forschung und Technologie, and SANOFI WINTHROP GmbH, Munich. The skillful assistance of Ms. Szasz (drawings) is gratefully acknowledged.

References

Alheid GF, Heimer L, Switzer RC (1990) Basal ganglia. In: Paxinos G (ed) The human nervous system. Academic Press, New York, pp 483–582

Amaral DG, Price JL, Pitkänen A, Carmichael ST (1992) Anatomical organization of the primate amygdaloid complex. In: Aggleton JP (ed) The amygdala: neurobiological aspects of emotion, memory, and mental dysfunction. Wiley-Liss, New York, pp 1–66

Bancher C, Braak H, Fischer P, Jellinger KA (1993) Neuropathological staging of Alzheimer lesions and intellectual status in Alzheimer's and Parkinson's disease. Neurosci Lett 162: 179–182

Braak H, Braak E (1986) Nuclear configuration and neuronal types of the nucleus niger in the brain of the human adult. Hum Neurobiol 5: 71–82

Braak H, Braak E (1989) Cortical and subcortical argyrophilic grains characterize a disease associated with adult onset dementia. Neuropathol Appl Neurobiol 15: 13–26

Braak H, Braak E (1990) Cognitive impairment in Parkinson's disease: amyloid plaques, neurofibrillary tangles and neuropil threads in the cerebral cortex. J Neural Transm [P D Sect] 2: 45–57

Braak H, Braak E (1991) Neuropathological stageing of Alzheimer-related changes. Acta Neuropathol 82: 239–259

Braak H, Braak E (1993) Anatomy of the human basal ganglia. In: Szelenyi I (ed) Inhibitors of monoamine oxidase B. Birkhäuser, Basel, pp 3–23

Braak H, Braak E (1994) Pathology of Alzheimer's disease. In: Calne BD (ed) Neurodegenerative diseases. Saunders, Philadelphia, pp 585–61

Braak H, Duyckaerts C, Braak E, Piette F (1993) Neuropathological staging of Alzheimer-related changes correlates with psychometrically assessed intellectual status. In: Corian B, Iqbal K, Nicolini M, Winblad B, Wisniewski H, Zatta PF (eds) Third International Conference on Alzheimer's Disease and Related Disorders. Chichester, Wiley, pp 131–137

Braak E, Braak H, Mandelkow EM (1994a) A sequence of cytoskeleton changes related to the formation of neurofibrillary tangles and neuropil threads. Acta Neuropathol 87: 554–567

Braak H, Braak E, Yilmazer D, de Vos RAI, Jansen ENH, Bohl J, Jellinger K (1994b) Amygdala pathology in Parkinson's disease. Acta Neuropathol 88: 493–500

Braak H, Braak E, Yilmazer D, Schultz C, de Vos RAI, Jansen ENH (1995) Nigral and extranigral pathology in Parkinson's disease. J Neural Transm [Suppl] 46: 15–31

Braak H, Braak E, Yilmazer D, de Vos RAI, Jansen ENH, Bohl J (1996) Pattern of brain destruction in Parkinson's and Alzheimer's disease. J Neural Transm 103: 455–490

Fearnley J, Lees A (1994) Pathology of Parkinson's disease. In: Calne DB (ed) Neurodegenerative diseases. Saunders, Philadelphia, pp 545–554

Felleman DJ, van Essen DC (1991) Distributed hierarchial processing in the primate cerebral cortex. Cerebral Cortex 1: 1–47

Gibb WRG (1991) Neuropathology of substantia nigra. Eur Neurol 31 [Suppl 1]: 48–59

Heimer L, Switzer RC, Van Hoesen GW (1982) Ventral striatum and ventral pallidum. Components of the motor system? Trends Neurosci 5: 83–87

Jellinger K (1990) New developments in the pathology of Parkinson's disease. In: Streifler MB, Korczyn AD, Melamed E, Youdim MBH (eds) Parkinson's disease: anatomy, pathology, and therapy. Raven Press, New York, pp 1–16 (Adv Neurol 53)

Jellinger K (1991) Pathology of Parkinson's disease. Changes other than the nigrostriatal pathway. Mol Chem Neuropathol 14: 153–197

Jellinger K (1994) Structural basis of dementia in Parkinson's disease. In: Korczyn AD (ed) Dementia in Parkinson's disease. Monduzzi Editore, Bologna, pp 31–38

Jellinger K, Bancher C (1995) Structural basis of mental impairment in Parkinson's disease. Neuropsychiatrie 9: 9–14

Jellinger K, Braak H, Braak E, Fischer P (1991) Alzheimer lesions in the entorhinal region and isocortex in Parkinson's and Alzheimer's diseases. Ann NY Acad Sci 640: 203–209

Lowe J (1994) Lewy bodies. In: Calne DB (ed) Neurodegenerative diseases. Saunders, Philadelphia, pp 51–69

Authors' address: Prof. Dr. med. H. Braak, Department of Anatomy, Theodor Stern Kai 7, D-60590 Frankfurt/Main, Federal Republic of Germany.

J Neural Transm (1996) [Suppl] 48: 7–21

New horizons in molecular mechanisms underlying Parkinson's disease and in our understanding of the neuroprotective effects of selegiline

M. Gerlach[1,2], **H. Desser**[3], **M. B. H. Youdim**[4], and **P. Riederer**[2]

[1] Neurologische Klinik, Ruhr-Universität Bochum, and
[2] Klinische Neurochemie, Universitäts-Nervenklinik, Würzburg, Federal Republic of Germany
[3] Ludwig-Boltzmann-Institut für Hämatologie, Abteilung Biochemie, Vienna, Austria
[4] Department of Pharmacology, Faculty of Medicine, Technion, Haifa, Israel

Summary. There have been many claims that the selective monoamine oxidase type B (MAO-B) inhibitor selegiline may have distinct properties in slowing the progression of Parkinson's disease (PD). Degeneration of nigrostriatal dopaminergic neurons is the primary histopathological feature of PD. Although many different hypotheses have been advanced, the cause of chronic nigral cell death and the underlying mechanisms remain elusive as yet. Therefore, there is no clear knowledge regarding an understanding of the reported effects of selegiline on the progression of PD. However, there is a considerable body of indirect evidence that oxidative stress may play a role in the pathogenesis of this illness. Oxidative stress refers to cytotoxic consequences of hydrogen peroxide and oxygen-derived free radicals such as the hydroxyl radical ($^{\cdot}OH$), the superoxide anion ($^{\cdot}O_2$), and nitric oxide (NO), which are generated as byproducts of normal and aberrant metabolic processes that utilize molecular oxygen. On the other hand, an increasing body of experimental data has implicated excitotoxicity as a mechanism of cell death in both acute and chronic neurological diseases. One of the receptor which is particularly involved in the toxic effects of excitatory amino acids is the NMDA (N-methyl-D-aspartate) receptor. Excessive stimulation of this type of receptor by glutamic acid or NMDA agonists leads to a massive influx of calcium ions into the neuron followed by activation of a variety of calcium-dependent enzymes, impaired mitochondrial function, and the generation of free radicals.

This article will consider the concept that excitotoxicity is linked with the generation of free radicals. In view of this idea it will be further discussed how selegiline might exert its neuroprotective effects via indirect actions on the polyamine binding site of the NMDA receptor. Under treatment with the MAO-B inhibitor selegiline, the degradation of putrescine via MAO, a key factor in regulating the polyamine metabolism, might be diminished in the Parkinsonian brain, which in turn would suppress the polyamine synthesis.

Hence, the reported neuroprotective effect of selegiline might also receive a contribution from the diminished potentiation of the NMDA receptor by the polyamine binding site. On the other hand, since N^1-acetylated spermine and spermidine are also good substrates of MAO-B, it is likely that these compounds will be present in the brain in increased concentrations. It therefore seems possible that they will exert a neuroprotective effect via an antagonistic modulation of the polyamine binding site of the NMDA receptor.

Introduction

Parkinson's disease (PD) is characterized by a gradual degeneration of neurons that synthesize the neurotransmitter dopamine within the substantia nigra pars compacta of the brain. The resultant decrease of dopamine-containing fibers within the striatum is associated with the loss of dopamine. This has been thought to be the major pathochemical correlate of the main symptoms of PD such as akinesia and rigidity (Bernheimer et al., 1973) and is the rationale for the dopamine-substitution therapy including treatment with L-DOPA (L-3,4-dihydroxyphenylalanine, levodopa) and a peripheral decarboxylase inhibitor (carbidopa or benserazide), dopaminergic agonists (bromocriptine, dihydro-α-ergocryptine, lisuride, pergolide) and the selective monoamine oxidase type B (MAO-B) inhibitor selegiline (L-deprenyl).

Based on its blockade of MAO-B, and the assumed increase of dopamine, selegiline improves the efficacy of the dopamine-substitution therapy with L-DOPA (for a review, see Wessel, 1993). In addition, it has been claimed that treatment with this MAO-B inhibitor has a dramatic effect on slowing the progression of PD. In the multi-centre DATATOP-study (L-deprenyl and tocopherol antioxidative therapy of Parkinsonism), the effect of α-tocopherol and selegiline was investigated in 800 patients with PD (The Parkinson Study Group, 1989, 1993). The rationale for these studies is based on the neuroprotectice properties of this substance in experimental models (for a review, see Gerlach et al., 1992) and on a retrospective study showing that patients who received both L-DOPA and selegiline lived longer than patients who were treated with L-DOPA alone (Birkmayer et al., 1985). An interim analysis of the DATATOP-study after 12 ± 5 months indicated that selegiline reduced the risk of disability requiring L-DOPA therapy by approximately 50% (The Parkinson Study Group, 1989). An extended analysis after a mean (±SD) follow-up of 14 ± 6 months showed that the beneficial effects of selegiline which occurred largely during the first 12 months of treatment, remained strong, and significantly delayed the onset of disability requiring L-DOPA therapy (The Parkinson Study Group, 1993). The difference in the estimated median time to the end-point was about nine months. However, it was unclear whether the effects of L-deprenyl would be sustained or whether selegiline resulted in only short-term amelioration of clinical features (symptomatic effect), a slowing of underlying nigral degeneration (protective effect), or both these mechanisms.

Similar results to those of the DATATOP-study were recently also reported by Przuntek (1994). This was similarly a placebo-controlled and multicentre longitudinal study designed to measure the effect of selegiline on the time course of changes in the required dosage of L-DOPA and of the L-DOPA-determined fluctuations in mobility (SELEDO-study: selegiline and L-DOPA long-period trial). Interim analysis of the findings after four years showed that L-DOPA dosage necessary for the treatment of Parkinsonian patients could be kept constant if the patients were treated in conjunction with selegiline. In addition, it was found fewer motor fluctuations and a retardation of the progression of PD with the early combination therapy of L-DOPA/selegiline compared to L-DOPA monotherapy (Przuntek, 1994).

Although many different hypotheses have been advanced, the cause of chronic nigral cell death and the underlying mechanisms remain elusive as yet. Therefore, there is no clear knowledge regarding an understanding of the reported effects of selegiline on the progression of PD. However, the partial elucidation of the processes which underlie the selective action of neurotoxic substances such as 6-hydroxydopamine (6-OHDA), glutamic acid (glutamate), kainic acid, quinolinic acid or 1-methyl-4-phenyl-1,2,3,6-tetrahydropyridine (MPTP), has revealed possible molecular mechanisms that give rise to illnesses (Table 1). Evidence from a number of studies in experimental animals has shown that selegiline causes protection against the damaging effects of several neurotoxins, including the dopaminergic agents MPTP and 6-OHDA (for a review, see Gerlach et al., 1992). Accordingly, possible mechanisms responsible for the observed neuroprotective effects have been suggested. The purpose of this article is to consider the excitotoxic hypothesis as a cause of nigral cell death in PD and how selegiline might exert its neuroprotective effects considering this concept.

The excitotoxic concept

The concept of excitotoxic cell death postulated by Olney (1978) implies two paradoxical modes of action of amino acid neurotransmitters (e.g. glutamic acid, aspartic acid) or of excitatory amino acid receptor agonists (e.g. kainic acid, quinolinic acid), namely that they exert a excitatory action at physiological concentrations while they cause a neurotoxic effect at higher concentrations. Excitatory amino acids bind to specific receptor binding sites, and operate to stimulate neurons by opening receptor-coupled ion channels (for a review, see Headley and Grillner, 1990). This action is essential for the normal functional activity of the CNS. In the case of a neurotoxic effect, there is a massive increase of the intracellular calcium concentration as a result of the excessive stimulation of glutamatergic neurons. The concept of excitotoxic cell death (for a review, see Olney, 1989, and Coyle and Puttfarcken, 1993), which has in recent years received experimental verification, points to possible mechanisms of how endogenous transmitters (glutamic acid for example), but also exogenous neurotoxins (for example β-N-methyl-amino-L-alanine, BMAA) can damage neurons.

Table 1. Possible molecular mechanisms of nigral cell death in Parkinson's disease (modified according to Gerlach et al., 1996)

Molecular mechanisms	Possible causative processes
Oxidative stress	
Neurotoxic effects of oxygen-derived free radicals	Hydrogen peroxide-producing intermediate reactions
	Metabolism of dopamine and endogenous or exogenous neurotoxins (e.g. MPTP or MPTP-resembling compounds)
	Impaired free-radical scavenging systems
	Altered brain iron metabolism
	Inflammatory cytokines-induced gliosis
Excitotoxic mechanism	
Excitatory amino acid receptor-mediated influx of cations gives rise to neurotoxic effects	Abnormal glutamate accumulation
	Exogenous excitotoxins (e.g. domoic acid, BMAA, the proposed exogenous excitotoxin in Guam disease)
Disturbance in mitochondrial energy metabolism	
Diminished or completely ceased ATP synthesis	Mitochondrial toxins (MPP^+, paraquat, nitric oxide)
	Glutamate-induced processes
	Disturbance of calcium homeostasis
Disturbance of intracellular calcium homeostasis	
Excessive activation of calcium-dependent enzymes (e.g. protein kinase C, phospholipases, proteases, endonucleases, nitric oxide synthase) that are involved in neuronal function	Cell membrane damages
	Impaired energy supply (e.g. inhibition of mitochondrial respiratory chain enzymes)
	Excitotoxins

BMAA β-N-methyl-amino-L-alanine; *MPTP* 1-methyl-4-phenyl-1,2,3,6-tetrahydropyridine; MPP^+ 1-methyl-4-phenylpyridinium cation

Two main subtypes of glutamate receptors have been recognized on the basis of their molecular cloning, electrophysiologic properties, and pharmacologic antagonism (for a review, see Lodge and Collingridge, 1990). The two main divisions are ionotropic (receptor that are coupled directly to membrane ion channels) and metabotropic (receptors that are coupled to G proteins and modulate intracellular second messengers such as inositol triphosphate, calcium, and cyclic nucleotides). An important point concerning these subtypes of glutamate receptors is that NMDA (N-methyl-D-aspartate) receptor-activated channels permit the influx of calcium as well as sodium (Fig. 1), and overstimulation of this type of receptor is one mechanism for calcium overload in neurons. In addition, some variants of AMPA (α-amino-3-hydroxy-5-methyl-4-isoxazolepropionic acid) and kainate receptors are coupled to ion channels that are somewhat permeable to calcium and can thus contribute to excessive calcium entry. Furthermore, stimulation of any of the ionotropic glutamate receptors results in membrane depolarisation because of the influx of positively charged ions, and thus indirectly activates voltage-gated calcium channels. The influx of calcium by way of these voltage-dependent channels may also contribute to glutamate-mediated neurotoxicity. Increased cytoplasmic calcium can activate a variety of calcium-dependent enzymes (Mayer and Miller, 1990). This activation sets mechanisms in train which lead to the decline of the neuron (for a review, see Coyle and Puttfarcken, 1993, and Gerlach et al., 1996). For example, the activation of calpain I and II leads to alterations in the cytoskeleton; the activation of protein-kinase C and nitric oxide synthase results in the formation of toxic free radicals; the activation of phospholipase A_2 leads to the breakdown of phospholipid membranes. The fatty acids liberated by this process, such as arachidonic acid, then enters the extracellular space and are there broken down further into free radicals. In this way a vicious circle of cell-damage is maintained or even reinforced. In this connection it is interesting to note the recently described finding according to which glutamate generated by astrocytes cultured from embryonic rat or mouse striatum is capable of releasing arachidonic acid (Stella et al., 1994).

Possible role of polyamines in excitotoxic mechanisms

The major cellular polyamines, putrescine (1,4-butanediamine, although putrescine is a diamine, for the sake of simplicity, putrescine will be referred to as a polyamine), spermidine (N-(3-aminopropyl)-1,4-butanediamine), and spermine (N, N-'bis(3-aminopropyl)-1,4-butanediamine) are polyvalent cations that are ubiquitiously distributed in mammalian cells (Pegg and McCann, 1988). Polyamines have multiple functions within the CNS (reviewed in Morrison et al., 1995), including roles in brain development, nerve growth and regeneration, response to brain injury and stress, brain metabolism, regulation of ionic flux and neuronal ion channels, and modulation of several neurotransmitter receptors in the brain. One of these receptors is the NMDA receptor (Fig. 1), whose activation is also modulated by polyamines, such as spermine and spermidine (Williams et al., 1991). Although the polyamine binding site needs not be occupied for receptor activation, the presence of polyamines increases the ability of glutamate and glycine to open the NMDA-receptor ion channel. It is therefore possible that polyamines potentiate excitotoxicity via NMDA receptor activation. Accordingly, antagonists acting at the polyamine binding site of the NMDA receptor should be effective as blockers of NMDA-induced neurotoxicity. Indeed, such a protective effect have been shown recently in murine cultured neocortical neurons (Beart et al., 1995).

In the adult brain, concentrations of spermidine were highest, followed by spermine and putrescine (Table 2). As shown in Table 2, there seems to be an uneven regional distribution of these polyamines in the brain. Indeed, a recent comprehensive investigation have demonstrated a distinct and uneven distribution pattern among ten examined brain areas (Morrison et al., 1995): Spermidine levels were especially high in the white matter and the thalamus (20 and 9.3 nmol/mg of protein, respectively), whereas spermine concentra-

NMDA Receptor

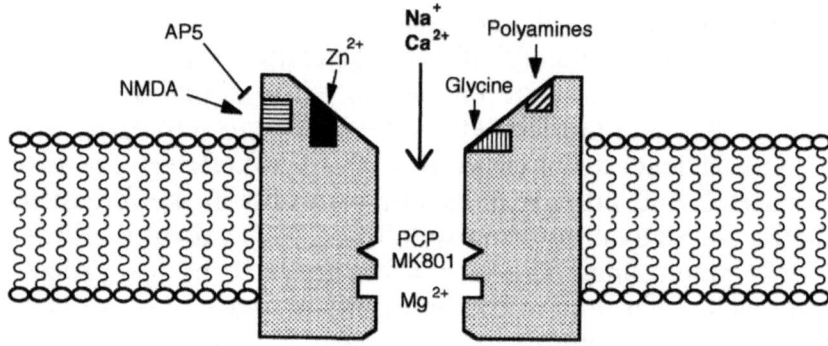

Fig. 1. Features of the N-methyl-D-aspartate (NMDA) receptor complex. The NMDA receptor gates a cation channel that is permeable to Ca^{2+} and Na^+ and is gated by Mg^{2+} in a voltage-dependent fashion; K^+ is the counterion. In addition to a binding site for glutamate, there is also a binding site for glycine. Both the glutamate and glycine binding sites must be occupied for receptor activation to occur. NMDA receptor activation is also modulated by polyamines, such as spermine and spermidine. Although the polyamine binding site needs not be occupied for receptor activation, the presence of polyamines increases the ability of glutamate and glycine to open the NMDA receptor ion channel. The NMDA receptor channel is blocked by phencyclidine (PCP) and dizocilpine maleate (MK801); 2-Amino-5-phosphonopentanoate (AP5) is an antagonist

Table 2. Distribution of polyamines in the human brain (Desser and Riederer, 1984)

Region	Putrescine	Spermidine	Spermine
Thalamus	9.1 ± 3.3 (3)	527 ± 315 (3)	95.6 ± 34.6 (3)
Brain stem	23.8 ± 23.0 (12)	325 ± 194 (11)	117.0 ± 72.7 (13)
Striatum	18.8 ± 35.5 (13)	403 ± 218 (15)	67.3 ± 30.1 (14)

Autopsied brain was obtained from subjects who died without evidence of neurological or psychiatrc disease or brain pathology. Polyamine levels were determined by reversed-phase HPLC with fluorescence detection, employing a pre-column derivatization method. The concentrations of polyamines were calculated as nmol per g wet weight of tissue. Data shown are mean polyamine levels ± SD obtained from n tissue samples

tions were highest in the cerebellar cortex (3.4 nmol/mg of protein); high levels of putrescine were observed in cerebral cortices, the putamen, and the hippocampus (0.7–1.2 nmol/mg of protein), with lowest levels in the cerebellum and the thalamus (0.3–0.5 nmol/mg of protein).

Using binding of the phencyclidine receptor ligand [^{3}H]MK801 as an in vitro probe of NMDA-channel activation, the mean half-maximal enhancement (EC_{50}) value for stimulation by spermidine was 13.3 ± 2.2 μM (Nussenzveig et al., 1991). Although the effective extracellular concentration of polyamines is unknown and may be several orders of magnitude lower than the found levels in brain homogenates, the mutually facilitatory effect of L-glutamate and sub-saturating concentration of spermidine may therefore be of physiological relevance. On the other hand, this effect may be of importance in excitotoxic injury due to induction of the polyamine system in response to a variety of stimuli, such as growth factors, tumor promoters, cerebral ischemia, mechanical and thermal brain injury, seizure activity, neurotoxin insult, and neuronal deafferentation (reviewed in Morrison et al., 1995).

Link between excitotoxicity and oxidative stress

In cultured cells it is possible to prevent the toxic action of excitatory amino acids by the administration of non-competitive NMDA receptor antagonists and by calcium channel blockers, which inhibit the entry of calcium ions (e.g. Azmitia, 1989; Erdö and Schäfer, 1991). Studies in cerebellar neurons have further shown, that several free radical scavengers (e.g. α-tocopherol, a biologically active component of vitamin E, ascorbic acid, ubiquinone, or 21-amino steroids) can attenuate glutamate toxicity (e.g. Majewska and Bell, 1990; Puttfarcken et al., 1993), whereas glutathione (GSH) depletion exacerbates toxicity (Bridges et al., 1991). In addition, cultured cortical neurons from mice overexpressing the superoxide radical scavenging enzyme superoxide dismutase (SOD) are less vulnerable to glutamate toxicity (Chan et al., 1990). These results implicate a link between excitotoxicity and oxidative stress in vitro.

The oxidative stress hypothesis infers an imbalance between the formation of cellular oxidants and the antioxidative processes. Oxidative stress refers to cytotoxic consequences of hydrogen peroxide and oxygen-derived free radicals such as the hydroxyl radical ($^{\cdot}OH$), the superoxide anion ($^{\cdot}O_2$), and nitric oxide (NO), which are generated as byproducts of normal and aberrant metabolic processes that utilize molecular oxygen (Halliwell, 1992; Götz et al., 1994). Hydrogen peroxide is produced in human tissues by several enzymes, such as SOD, L-amino acid oxidase, glycollate oxidase, xanthine oxidase, and MAO (Halliwell, 1992). In dopaminergic nerve cells it is mainly generated by MAO via deamination of dopamine, and non-enzymatically by autoxidation of dopamine. Hydrogen peroxide is relatively inert and not toxic to cells (Halliwell, 1992). However, damage is done when hydrogen peroxide interacts with the reduced forms of transitional metal ions (e.g. iron (II) or copper (I)) and decomposes to the highly reactive hydroxyl radical (the Fenton reaction). In addition, hydroxyl radicals are produced in the mitochondria of neurons during oxidative phosphorylation, as shown in equation 1 below.

$$O_2 \xrightarrow[]{+e^-} {}^{\cdot}O_2^- \xrightarrow[+2H^+]{+e^-} H_2O_2 \xrightarrow[]{+e^-} {}^{\cdot}OH + OH^- \xrightarrow[+2H^+]{+e^-} 2H_2O \qquad (1)$$

Hydroxyl radicals rapidly react with and have a strong affinity for almost every molecular species found in living cells. Such reactions include breakage of single- and double-stranded DNA, chemical alterations of the deoxyribose purine and pyrimidine bases, membrane lipids and carbohydrates (Halliwell, 1992), leading to a cascade of events with subsequent damage to the mitochondial electron-transport system, decompartmentalization of intracellular calcium homeostasis, induction of proteolysis by proteases, increased membrane lipid peroxidation, and finally cell death (Götz et al., 1994; Gerlach et al., 1996).

Although the proof that oxidative stress actually causes the loss of dopaminergic neurons in patients with PD is lacking, there is a considerable body of indirect evidence from experimental models of PD and post mortem studies in PD that supports this concept (for a review, see Fahn and Cohen, 1992; Lange et al., 1992; Gerlach and Riederer, 1993): In the brains of patients who have died with PD, specifically the substantia nigra has been found to contain diminished activities of glutathione peroxidase and catalase (hydrogen peroxide metabolizing enzymes), as well as reduced concentrations of GSH. These findings suggest an aberrant metabolism of hydrogen peroxide. There is also a site-specific increase of SOD activity in the substantia nigra (e.g. Saggu et al., 1989) as well as a slightly increased MAO-B activity (+25% compared to the values of individuals who during life were not diagnosed to have any previous neurological or psychiatric diseases) (Riederer et al., 1989). Both of these findings point to an increased hydrogen peroxide formation in the substantia nigra of patients with PD. The increase of iron in the substantia nigra (Sofic et al., 1988; Dexter et al., 1989b) with a shift of the nigral iron(II)/iron(III) ratio from 2:1 in control brains to 1:2 in the brains of patients with PD (Sofic et al., 1988), indicates an increased rate of synthesis of hydroxyl radicals. Necropsy studies have shown increased basal levels of thiobarbituric acid-reactive substances in the substantia nigra of patients with PD (a measure of secondary products of lipid peroxidation) coupled with a decrease in the levels of polyunsaturated fatty acids (the substrates for lipid peroxidation) (Dexter et al., 1989a). In addition, there seems to be radical-induced DNA damage as indicated by raised 8-hydroxydeoxyguanosine (Sanchez-Ramos et al., 1994), a product of free-radical attack on guanine in DNA.

Further evidence for the occurrence of oxidative stress in PD comes from studies on experimental models of this disease. For example, the iron chelator desferrioxamine (desferal) and α-tocopherol protect rats against the 6-OHDA-induced reduction in striatal dopamine content and decrease of dopamine-related spontaneous locomotor activity (Ben-Shachar et al., 1991). These findings indicate the prevention of 6-OHDA-induced degeneration of nigro-striatal dopaminergic neurons. 6-OHDA is thought to induce nigro-striatal dopaminergic lesions via generation of hydrogen peroxide and hydroxyl radicals derived from it, presumably initiated by transition metal such as iron. In fact, it has been shown by magnetic resonance (MR) imaging (Hall et al., 1992), and by neurochemical and histochemical studies (e.g. Oestreicher et al., 1994) that iron is increased in the striatum of 6-OHDA-lesioned rats. Furthermore it has been shown, that 6-OHDA releases iron from ferritin in vitro (Monteiro and Winterbourn, 1989). Finally, intranigral injections of iron(III) produce neurotoxic effects similar to those observed with 6-OHDA

(Ben-Shachar and Youdim, 1991), and leads to a progressive reduction of striatal dopamine metabolism after a single intranigral administration (Wesemann et al., 1994).

Although earlier data showed that MPTP-induced dopaminergic neurotoxicity does not involve oxidative stress (for a review, see Gerlach et al., 1991), it was shown in a previous study that the neurotoxic action of MPTP involves the generation of hydroxyl radicals from released dopamine (Chiueh et al., 1993). In addition it has been demonstrated that hydrogen peroxide and hydroxyl radicals are also products of the interaction of 1-methyl-4-phenylpyridinium cation (MPP$^+$, the probable toxic metabolite of MPTP, which is accumulated intraneuronally by the high-affinity dopamine-uptake system) with mitochondrial NADH dehydrogenase (Adams et al., 1993). The neurotoxic effects of MPTP (induction of a Parkinson-like akinesia in monkeys, which is accompanied by a striatal dopamine deficiency and a diminished cell number of neurons showing tyrosine hydroxylase immunoreactivity) can be completely prevented by prior treatment with the selective MAO-B inhibitor selegiline (for a review, see Gerlach et al., 1992) which prevents the metabolism of MPTP to the neurotoxic MPP$^+$. Mice that are transgenic for and overexpress human copper/zinc SOD are much less vulnerable to MPTP than control mice (Przedborski et al., 1992). This result also suggests that radical-mediated mechanisms are involved in the neurotoxic effects of MPTP. However, the neuroprotective actions of antioxidants such as ascorbic acid and α-tocopherol against MPTP toxicity are uncertain (Gerlach et al., 1991).

The most direct evidence that excitotoxicity and oxidative stress may be sequential and interactive mechanisms leading to neuronal degeneration is the finding that NMDA exposure leads to superoxide generation in cultures of cerebellar neurons (Lafon-Cazal et al., 1993). In a more recent investigation it was shown that free radical spin traps, such as α-phenyl-N-tert-butylnitrone (PBN) and N-tert-butyl-α-(2-sulfophenyl)-nitrone (S-PBN), can attenuate excitotoxic lesions in vivo (Schulz et al., 1995). These compounds react with unstable free radicals to produce more stable nitroxides. Pretreatment with S-PBN significantly attenuated striatal excitotoxic lesions in rat produced by NMDA, kainic acid, and AMPA. In a similar manner, striatal lesions produced by MPP$^+$, malonate, and 3-acetylpyridine were significantly attenuated by either S-PBN or PBN treatment. These results provide an in vivo evidence for the involvement of free radicals in excitotoxicity.

Excitotoxic injury of nigral neurons as a cause of Parkinson's disease

As discussed previously, evidence is now emerging that activation of glutamate-gated cation channels may be an important source of oxidative stress and that these two mechanisms act in a sequential as well as a reinforcing manner, leading to selective neuronal degeneration. Further confirmation for a link between excitotoxicity and oxidative stress is suggested by the finding that the non-competitive NMDA antagonist MK801 (dizocilpine

maleate) (Zuddas et al., 1992) and the competitive NMDA antagonist CPP
(($\pm$)-2-carboxypiperazine-4-yl-propyl-1-phosphonic acid) (Lange et al., 1993)
are able to protect against MPTP-induced Parkinsonism in primates, although
findings concerning the role played by excitotoxicity on MPTP toxicity in
rodents are controversial (e.g. Kupsch et al., 1992).

However, a prerequisite that these interacting mechanisms are involved in
the pathogenesis of PD is that glutamate-gated ion channels must be located
on either nigrostriatal nerve endings or nigral dendrites. Indeed, recent data
from investigations utilizing extracellular single-unit recordings in chloral-
anesthetized rats have indicated that both NMDA and AMPA receptors are
present on substantia nigra dopamine neurons (Christoffersen and Meltzer,
1995). Autoradiographic studies have identified high levels of NMDA and
AMPA receptors in the striatum of the rat (Albin et al., 1992). However,
although it has been suggested that there are pre-synaptic excitatory amino
acid receptors in the striatum, the majority of striatal excitatory amino acid
binding sites appear to be localized post-synaptically on striatal neurons
(Greenamyre and Young, 1989). Hence, glutamate-gated cation channels
must be located on nigral dendrites, if the cause of nigral dopaminergic cell
death is due to the previously discussed interacting mechanisms. Indeed, in
autoradiographic image studies there have been observed the binding of all
excitatory amino acid receptor subtypes in the human substantia nigra
(Difazio et al., 1992). However, the binding of all excitatory amino acid
receptor subtypes in the midbrain was quite low, being approximately one-
fiftieth that found in the hippocampus. Despite the low levels of glutamate
binding, there was a significant reduction of approximately 80% in NMDA-
sensitive [^{3}H]glutamate binding in the substantia nigra pars compacta, suggesting that nigral NMDA binding sites are concentrated on these neurons
(Difazio et al., 1992).

Indirect effect of selegiline on the function of the NMDA receptor

The foregoing discussion implies that excitotoxicity and oxidative stress may
be sequential and interactive mechanisms leading to neuronal cell degeneration in PD. Therefore, the apparent neuroprotective effects of selegiline might
be understood considering its indirect actions on the polyamine site of the
NMDA receptor.

As discused previously, polyamines modulate the response of the NMDA
receptor through a unique allosteric regulatory site: Spermine and spermidine
potentiate the response of the NMDA receptor, while putrescine is a weak
competitive inhibitor of spermine activation (McGurk et al., 1990;
Nussenzveig et al., 1991). As might be expected from the regulatory nature of
these compounds, polyamine synthesis is a tightly controlled process (see Fig.
2). This permits subtle adjustment of intracellular concentrations according to
physiological needs. Ornithine is the exclusive precursor of putrescine in the
vertebrate organism, from which it is formed by enzymatic decarboxylation
through ornithine decarboxylase (Fig. 2). S-Adenosylmethionine (SAM) de-

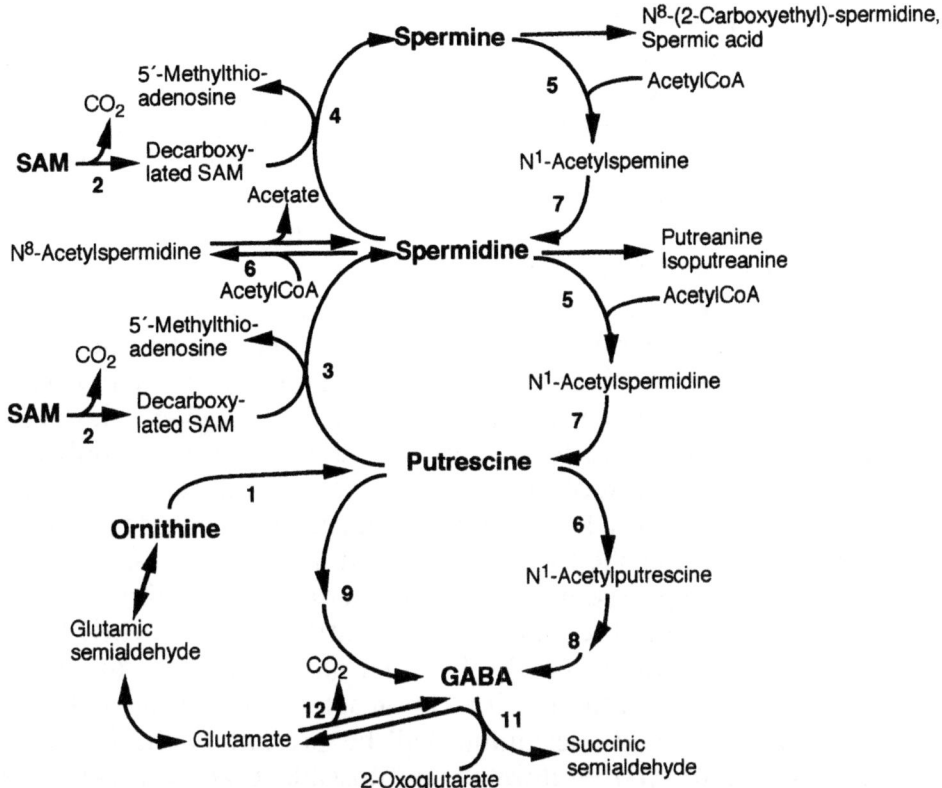

Fig. 2. Polyamine metabolism (according to Seiler, 1981). *SAM* S-adenosylmethionine, *GABA* γ-aminobutyric acid. *1* ornithine decarboxylase; *2* S-adenosylmethionine decarboxylase; *3* spermidine synthase; *4* spermine synthase; *5* AcetylCoA: polyamine N^1-acetyltransferase (cytoplasmic); *6* AcetylCoA: spermidine N^8-acetyltransferase (nuclear); *7* polyamine oxidase; *8* monoamine oxidase; aldehyde dehydrogenase; *9* diamine oxidase; aldehyde dehydrogenase; *10* ornithine: 2-oxoacid aminotransferase; *11* GABA transaminase; *12* glutamic acid decarboxylase

carboxylase catalyzes the decarboxylation of SAM to form decarboxylated SAM, the donor of aminopropyl groups for spermidine and spermine synthesis. In addition, spermine may be metabolized back to spermidine, spermidine to putrescine, and putrescine to GABA through the polyamine interconversion pathway (Fig. 2). Thus, polyamine metabolism is a cyclic process which allows the transformation of putrescine into spermidine and spermine, and vice versa, according to physiological requirements. The metabolic cycle is essential for the regulation of polyamine turnover. The key-limiting enzyme of the interconversion pathway is spermidine/spermine N-acetyltransferase. All three of the above-mentioned enzymes are highly regulated, inducible enzymes with a high turnover rate (Seiler, 1981). Rapid induction of the rate-limiting biosynthetic decarboxylases and a high rate of inactivation permit spurts of polyamines to be produced. Putrescine and the decarboxylated SAM are at the center of complex regulating systems controlling the rate-limiting enzymes (Seiler, 1981).

In the adult brain with low diamine oxidase activity (Burckard et al., 1963) degradation of putrescine to GABA via monoacetylation to N^1-

acetylputrescine (Fig. 2) is the main catabolic pathway of polyamines (Seiler, 1981). Beside dopamine and phenylethylamine, monoacetylated polyamines are among the most important endogenous MAO-B substrates in the brain (Youdim, unpublished results). Because of the high activity of MAO-B in the liver (Kalaria and Harik, 1987) the N^1-acetylputrescine concentration is very low in this tissue (Seiler and Knödgen, 1980). In the human brain, the highest specific activity of MAO-B found is in the hypothalamus and in the nigro-striatal system (Riederer and Youdim, 1986). Hence, it is likely that in brain regions with high MAO-B activities, the polyamine metabolism is induced due to a high rate of putrescine degradation. Considering the finding of increased MAO-B activity in the substantia nigra of PD patients (Riederer et al., 1989), it is reasonable to assume that a higher catabolism of monoacetylated putrescine leads to an increased biosynthesis of polyamines with subsequent potentiation of the NMDA receptor-linked excitotoxicity.

Under treatment with the MAO-B inhibitor selegiline, the degradation of putrescine might be diminished which in turn would suppress the polyamine synthesis. Hence, the reported neuroprotective effect of selegiline might also receive a contribution from the diminished potentiation of the NMDA receptor by the polyamine binding site. On the other hand, regarding inhibition of MAO-B, as is the case in patients with PD on selegiline, it is also likely that N^1-acetylated spermine and spermidine will be present in the brain in increased concentrations. It therefore seems possible that they will exert a neuroprotective effect via an antagonistic modulation of the polyamine binding site of the NMDA receptor. But, because of the methodological difficulties associated with the method for the determination and the instability of such compounds, direct experimental evidence is lacking.

Acknowledgements

Research carried out in the laboratories of the authors (M.G. and P.R.) was generously supported by grants from the Bundesministerium für Forschung und Technologie (01 KL 9013; 01 KL 9101-0), and that of M.B.H.Y. by Golding Parkinson Disease Research Fund Technion, Haifa. The supports of the European Community in Brussels (Biomed I) and Fogarty International Center for Advanced Studies, N.I.H. Bethesda, are gratefully acknowledged.

References

Adams JD, Klaidman LK, Leung AC (1993) MPP+ and MPDP+ induced oxygen radical formation with mitochondrial enzymes. Free Rad Biol Med 15: 181–186

Albin RL, Makowiec RL, Hollingworth ZR, Dure LS, Penney Jr JB, Young AB (1992) Excitatory amino acid binding sites in the basal ganglia of the rat: a quantitative autoradiographic study. Neuroscience 46: 35–48

Azmitia EC (1989) Nimodipine attenuates NMDA- and MDMA-induced toxicity of cultured fetal serotonergic neurons: evidence for a generic model of calcium toxicity. In: Traber J, Gispen WH (eds) Nimodipine and central nervous system function: new vistas. Schattauer, Stuttgart New York, pp 141–159

Beart PM, Schousboe A, Frandsen A (1995) Blockade by polyamine NMDA antagonists related to ifenprodil of NMDA-induced synthesis of cylic GMP, increases in calcium and cytotoxicity in cultured neurones. Br J Pharmacol 114: 1359–1364

Ben-Shachar D, Youdim MBH (1991) Intranigral iron injection induces behavioral and biochemical "Parkinsonism" in rats. J Neurochem 57: 2133–2135

Ben-Shachar D, Eshel G, Finberg JPM, Youdim MBH (1991) The iron chelator desferrioxamine (desferal) retards 6-hydroxydopamine-induced degeneration of nigro-striatal dopamine neurons. J Neurochem 56: 1441–1444

Bernheimer H, Birkmayer W, Hornykiewicz O, Jellinger K, Seitelberger F (1973) Brain dopamine and the syndromes of Parkinson and Huntington. Clinical, morphological and neurochemical correlations. J Neurol Sci 20: 415–455

Birkmayer W, Knoll J, Riederer P, Youdim MBH, Hars V, Marton J (1985) Increased life expectancy resulting from addition of L-deprenyl to Madopar^R treatment in Parkinson's disease: a long-term study. J Neural Transm 64: 113–127

Bridges RJ, Koh J-Y, Hatalski CG, Cotman CW (1991) Increased excitotoxic vulnerability of cortical cultures with reduced levels of glutathione. Eur J Pharmacol 192: 199–220

Burckard WP, Gey KF, Pletscher A (1963) Diamine oxidase in polyamine catabolism. J Neurochem 10: 183–186

Chan PH, Chu L, Chen SF, Carlson EJ, Epstein CJ (1990) Reduced neurotoxicity in transgenic mice overexpressing human copper-zinc-superoxide dismutase. Stroke [Suppl III]21: III80–III82

Chiueh CC, Miyake H, Peng MT (1993) Role of dopamine autoxidation, hydroxyl radical generation and calcium overload in underlying mechanism involved in MPTP-induced parkinsonism. Adv Neurol 60: 251–258

Christoffersen CL, Meltzer LT (1995) Evidence for N-methyl-D-aspartate and AMPA subtypes of the glutamate receptor on substantia nigra dopamine neurons: possible preferential role for N-methyl-D-aspartate receptors. Neuroscience 67: 373–381

Coyle JT, Puttfarcken P (1993) Oxidative stress, glutamate, and neurodegenerative disorders. Science 262: 689–695

Desser H, Riederer P (1984) Polyamines in the human central nervous system: preliminary data in hepatic encephalopathy. International Conference on Polyamines, Budapest, August 6–10, 1984

Dexter DT, Cater CJ, Wells FR, Javoy-Agid F, Agid Y, Lees A, Jenner P, Marsden CD (1989a) Basal lipid peroxidation in substantia nigra is increased in Parkinson's disease. J Neurochem 52: 381–389

Dexter DT, Wells FR, Lees AJ, Agid F, Agid Y, Jenner P, Marsden CD (1989b) Increased nigral iron content and alterations in other metal ions occuring in brain in Parkinson's disease. J Neurochem 52: 1830–1836

Difazio MC, Hollingsworth Z, Young AB, Penney JB (1992) Glutamate receptors in the substantia nigra of Parkinson's disease brains. Neurology 42: 402–406

Erdö SL, Schäfer M (1991) Memantine is highly potent in protecting cortical cultures against excitotoxic cell death evoked by glutamate and N-methyl-D-aspartate. Eur J Pharmacol 198: 215–217

Fahn S, Cohen G (1992) The oxidant stress hypothesis in Parkinson's disease. Evidence supporting it. Ann Neurol 32: 804–812

Gerlach M, Riederer P (1993) The pathophysiological basis of Parkinson's disease. In: Szelenyi I (ed) Inhibitors of monoamine oxidase B. Birkhäuser, Basel Boston Berlin, pp 25–50

Gerlach M, Riederer P, Przuntek H, Youdim MBH (1991) MPTP mechanisms of neurotoxicity and their implications for Parkinson's disease. Eur J Pharmacol [Mol Pharmacol Sect] 208: 273–286

Gerlach M, Riederer P, Youdim MBH (1992) The molecular pharmacology of L-deprenyl. Eur J Pharmacol [Mol Pharmacol Sect] 226: 97–108

Gerlach M, Riederer P, Youdim MBH (1996) Molecular mechanisms for neurodegeneration: synergism between reactive oxygen species, calcium and excitotoxic amino acids. Adv Neurol 69: 177–194

Götz ME, Künig G, Riederer P, Youdim MBH (1994) Oxidative stress: free radical production in neural degeneration. Pharmacol Ther 63: 37–122

Greenamyre JT, Young AB (1989) Synaptic localization of striatal NMDA, quisqualate and kainate receptors. Neurosci Lett 101: 133–137

Hall S, Rulledge JH, Schallert T (1992) MRI brain iron and 6-hydroxydopamine experimental Parkinson's disease. J Neurol Sci 113: 198–208

Halliwell B (1992) Reactive oxygen species and the central nervous system. J Neurochem 59: 1609–1623

Headley PM, Grillner S (1990) Excitatory amino acid neurotoxicity and synaptic transmission: the evidence for a physiological function. In: Lodge D, Collingridge G (eds) The pharmacology of excitatory amino acids: a TiPS special report. Elsevier, Cambridge, pp 30–36

Kalaria RN, Harik S (1987) Blood-brain barrier monoamine oxidase. Enzyme characterization in cerebral microvessels and other tissues from six mammalian species, including human. J Neurochem 49: 856–864

Kupsch A, Loeschmann P, Sauer H, Arnold G, Renner P, Pufal D, Burg M, Wachtel H, ten Bruggencate G, Oertel WH (1992) Do NMDA receptor antagonists protect against MPTP-toxicity? Biochemical and immunocytochemical analyses in black mice. Brain Res 592: 74–83

Lafon-Cazal M, Pietri S, Culcasi M, Bockaert J (1993) NMDA-dependent superoxide production and neurotoxicity. Nature 364: 535–537

Lange KW, Youdim MBH, Riederer P (1992) Neurotoxicity and neuroprotection in Parkinson's disease. J Neural Transm [Suppl] 38: 27–44

Lange KW, Löschmann PA, Sofic E, Burg M, Horowski R, Kalveram KT, Wachtel H, Riederer P (1993) The competitive NMDA antagonist CPP protects substantia nigra neurons from MPTP-induced degeneration in primates. Naunyn Schmiedebergs Arch Pharmacol 348: 586–592

Lodge D, Collingridge G (1990) The pharmacology of excitatory amino acids: a TiPS special report. Elsevier, Cambridge

Majewska MD, Bell JA (1990) Ascorbic acid protects neurons from injury induced by glutamate and NMDA. NeuroReport 1: 194–196

Mayer ML, Miller RJ (1990) Excitatory amino acid receptors, second messengers and regulation of intracellular Ca^{2+} in mammalian neurons. In: Lodge D, Collingridge G (eds) The pharmacology of excitatory amino acids: a TiPS special report. Elsevier, Cambridge, pp 36–42

McGurk JF, Bennett MVL, Zukin RS (1990) Polyamines potentiate responses of N-methyl-D-aspartate receptors expressed in Xenopus oocytes. Proc Natl Acad Sci USA 87: 9971–9974

Monteiro HP, Winterbourn CC (1989) 6-Hydroxydopamine releases iron from ferritin and promotes ferritin-dependent lipid peroxidation. Biochem Pharmacol 38: 4177–4182

Morrison LD, Becker L, Ang LC, Kish SJ (1995) Polyamines in human brain: regional distribution and influence of ageing. J Neurochem 65: 636–642

Nussenzveig IZ, Sircar R, Wong M-L, Frusciante MJ, Javitt DC, Zukin SR (1991) Polyamine effects upon N-methyl-D-aspartate receptor functioning: differential alteration by glutamate and glycine site antagonists. Brain Res 561: 285–291

Oestreicher E, Sengstock GJ, Riederer P, Olanow CW, Dunn AJ, Arendash GW (1994) Degeneration of nigrostriatal dopaminergic neurons increases iron within the substantia nigra: a histochemical and neurochemical study. Brain Res 660: 8–18

Olney JW (1978) Neurotoxicity of excitatory amino acids. In: McGeer EG, Olney JW (eds) Kainic acid as a tool in neurobiology. Raven Press, New York, pp 95–121

Olney JW (1989) Excitatory amino acids and neuropsychiatric disorders. Biol Psychiatry 26: 505–525

Pegg AE, McCann PP (1988) Polyamine metabolism and function in mammalian cells and protozoans. ISI Atlas of Science (Biochemistry): 11–18

Przedborski S, Kostic V, Jackson LV, Naini AB, Simonetti S, Fahn S, Carlson E, Epstein CJ, Cadet JL (1992) Transgenic mice with increased Cu/Zn-superoxide dismutase activity are resistant to N-methyl-4-phenyl-1,2,3,6-tetrahydropyridine-induced neurotoxicity. J Neurosci 12: 1658–1667

Przuntek H (1994) Clinical aspects of neuroprotection in Parkinson's disease. J Neural Transm [Suppl] 43: 163–169

Puttfarcken PS, Getz RL, Coyle JT (1993) Kainic acid-induced lipid peroxidation: protection with butylated hydroxyl-toluene and U78517F in primary cultures of cerebellar granule cells. Brain Res 624: 223–232

Riederer P, Youdim MBH (1986) Monoamine oxidase activity and monoamine metabolism in brains of parkinsonian patients treated with L-deprenyl. J Neurochem 46: 1359–1365

Riederer P, Sofic E, Rausch WD, Hebenstreit G, Bruinvels J (1989) Pathobiochemistry of the extrapyramidal system: a "short note" review. In: Przuntek H, Riederer P (eds) Early diagnosis and preventive therapy in Parkinson's disease. Springer, Wien New York, pp 139–149 (Key Topics in Brain Research)

Saggu H, Cooksey J, Dexter D, Wells FR, Lees A, Jenner P, Marsden CD (1989) A selective increase in particulate superoxide dismutase activity in parkinsonian substantia nigra. J Neurochem 53: 692–697

Sanchez-Ramos JR, Övervik E, Ames BN (1994) A marker of oxyradical-mediated DNA damage (8-hydroxy-2′deoxyguanosine) is increased in nigro-striatum of Parkinson's disease brain. Neurodegeneration 3: 197–204

Schulz JB, Henshaw DR, Siwek D, Jenkins BG, Ferrante RJ, Cipolloni PB, Kowall NW, Rosen BR, Beal MF (1995) Involvement of free radicals in excitotoxicity in vivo. J Neurochem 64: 2239–2247

Seiler N (1981) Review. Polyamine metabolism and function in brain. Neurochem Int 3: 95–110

Seiler N, Knödgen B (1980) High-performance liquid chromatographic procedure for the simultaneous determination of the natural polyamines and their monoacetyl derivatives. J Chromatogr 221: 227–235

Sofic E, Riederer P, Heinsen H, Beckmann H, Reynolds GP, Hebenstreit G, Youdim MBH (1988) Increased iron(III) and total iron content in post mortem substantia nigra of parkinsonian brain. J Neural Transm 74: 199–205

Stella N, Tence M, Glowinski J, Premont D (1994) Glutamate-evoked release of arachidonic acid from mouse brain astrocytes. J Neurosci 14: 568–575

The Parkinson Study Group (1989) Effect of deprenyl on the progression of disability in early Parkinson's disease. N Engl J Med 321: 1364–1371

The Parkinson Study Group (1993) Effects of tocopherol and deprenyl on the progression of disability in early Parkinson's disease. N Engl J Med 328: 176–184

Wesemann W, Blaschke S, Solbach M, Grote C, Clement H-W, Riederer P (1994) Intranigral injected iron progressively reduces striatal dopamine metabolism. J Neural Transm [PD Sect] 8: 209–214

Wessel K (1993) MAO-B inhibitors in neurological disorders with special reference to selegiline. In: Szelenyi I (ed) Series of new drugs, vol 1. Inhibitors of monoamine oxidase B. Birkhäuser, Basel, pp 253–275

Williams K, Romano C, Dichter MA, Molinoff PB (1991) Modulation of the NMDA receptor by polyamines. Life Sci 48: 469–479

Zuddas A, Oberto G, Vaglini F, Fascetti F, Fornai F, Corsini GU (1992) MK-801 prevents 1-methyl-4-phenyl-1,2,3,6-tetrahydropyridine-induced parkinsonism in primates. J Neurochem 59: 733–739

Authors' address: PD Dr. M. Gerlach, Klinische Neurochemie, Universitäts-Nervenklinik, Füchsleinstrasse 15, D-97080 Würzburg, Federal Republic of Germany

J Neural Transm (1996) [Suppl] 48: 23–28

Early diagnosis in Parkinson's disease — limitation of biochemical markers and instrumental methods

P. H. Kraus

Department of Neurology, St. Josef Hospital, Bochum, Federal Republic of Germany

Summary. For most cases of Parkinson's disease (PD) we can estimate the time lag from the first onset of unspecific complaints until diagnosis by case history to be about 2 years. Until now the exact course of neurodegeneration in PD is still unknown. On base of recent knowledge we can discuss different models for the development of PD. There are substantially different possible approaches to recognise PD as soon as possible. The time interval for an earlier diagnosis is different for these methods and depends on the real course of neurodegeration.

At the time when the diagnosis of Parkinson's disease (PD) can be made with help of clinical and neuroimaging methods about 60 to 80% of nigral dopaminergic cells are lost. If we want to protect a relevant number of these cells, we have to recognise PD at a time, when still enough cells are intact.

The diagnosis of PD is established usually by physical examination and case history. For most cases we can estimate the time lag from the first onset of unspecific complaints until diagnosis to be about 2 years.

Some patients' history include a first temporary occurrence of symptoms under stress many years before symptoms appear again but remaining. Some authors estimate the presymptomatic phase of PD between 10 an 20 years.

We do not know the exact course of the degeneration and we until now do not know the causal chain of the underlying process although we know a few presumed causative parameters.

Until now we don't have any specific help for diagnosis during the preclinical phase.

In the following basic problems of an earlier diagnosis and advantages as well as disadvantages of different approaches will be presented and discussed.

Models for time course of degeneration

It is not even clear, whether the decline of the nerve cell number follows a certain (e.g. linear) curve, but in the following an exponential function has been chosen since this is the most common in nature and there are some data from animal models which support this estimation.

On base of recent knowledge we can discuss different models for the development of PD:

One possible reason for PD could be an initial decrease of the system before birth followed by normal ageing. Clinical symptoms will appear when over time a certain threshold is reached.

This model does not take into account the development of the basal ganglia during the first two decades of life. Taking into consideration cases with juvenile Parkinsonism the clinical picture is quite different in cases with very early beginning (Segawa, 1986).

In this model (where we don't know the extension of the initial defect) the period for development would be about 50 to 60 years (Fig. 1a, assumption of initial decline of about 40%).

After a possible environmental insult there also may be a time lag caused by following normal ageing (Fig. 1d). For this model we can find different examples like postencephalitic PD (where is a remarkable delay) and patients developing Parkinsonism after exposition to MPTP. Also after intoxication with carbon monoxide there are few cases with a (shorter) time lag.

Another model is that of accelerated ageing caused by an initial (unknown) defect. (Fig. 1b). If an acquired or late manifesting chronic process is the reason for PD, degeneration would follow a curve shape like Fig. 1c.

Nevertheless reality is much more complicated:

Also a combination of different models might be possible as well as different causes might lead to the same result.

The degenerative process might be also fluctuating over time.

Furthermore we don't know enough about compensating mechanisms of the central nervous system during long-time processes. If compensation plays an important role, we possibly can not differentiate the course from that of an environmental insult, when compensation breaks down at a certain level (Fig. 2). Most data which could give arguments for a certain model stem from cross sectional studies and can therefore not be easily interpreted as longitudinal.

Especially because of inter-individual differences cross sectional examinations are not able to catch any process which occurs as a step — events can be measured only by longitudinal studies which start before the step with repetitive examinations.

Our aim is to recognise PD as soon as possible. One basic problem of most approaches is that of specificity: For a number of parameters significant differences between healthy controls and PD patients could be shown. But this findings do not help very much for an earlier diagnosis in a certain patient. What we need is a 100% right classification of a certain case — the underlying question is not whether this patient is healthy or suffering from PD but whether the patient has PD or not (the latter part includes health as well as all relevant differential diagnoses).

There are substantially different approaches to an earlier diagnosis in PD (Table 1).

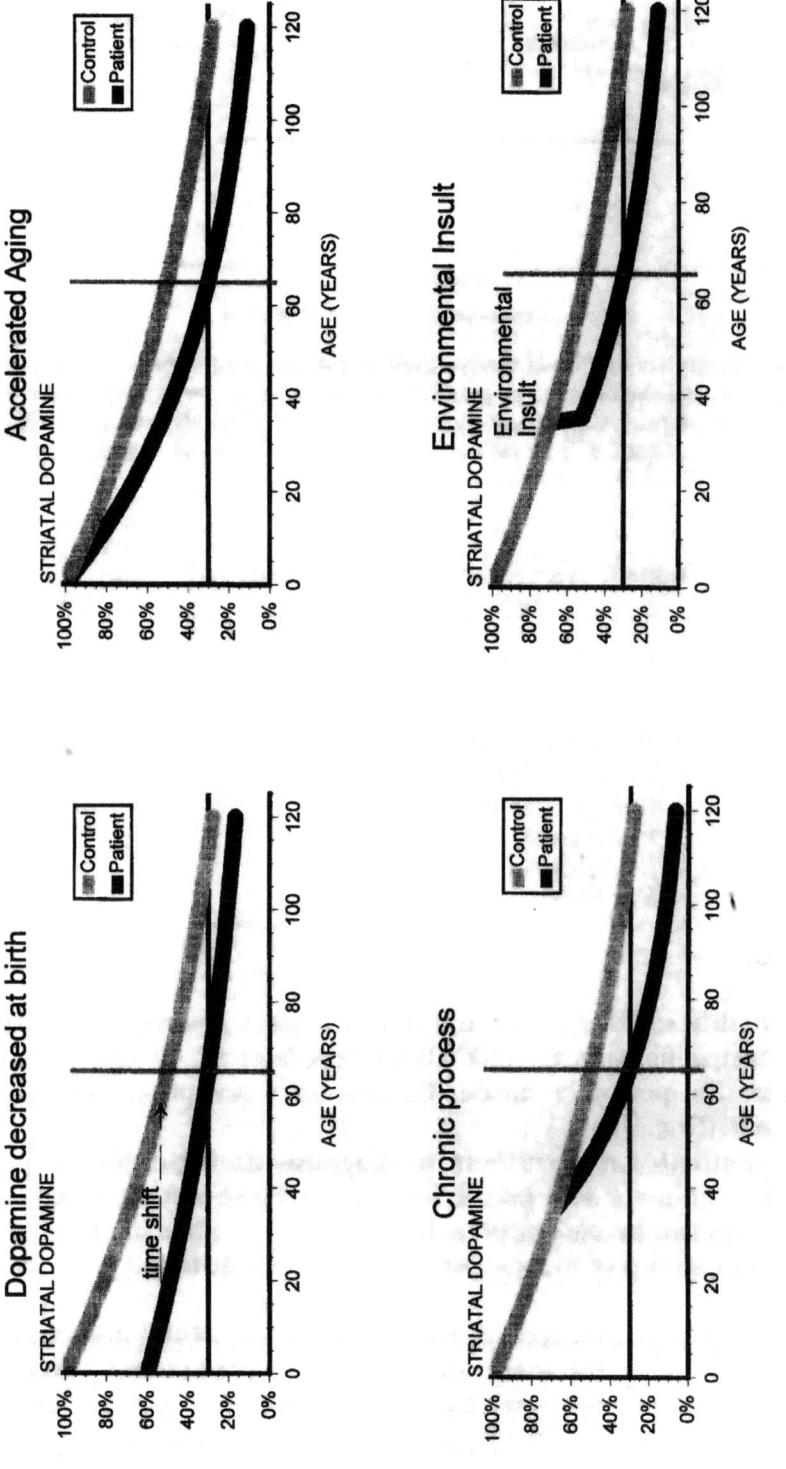

Fig. 1. Models of degeneration in PD: Four possible courses of degeneration in Parkinson's disease

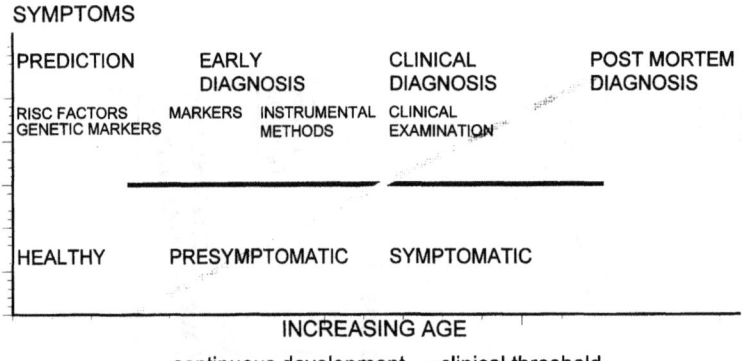

Fig. 2. Model of diagnosis in PD: If compensation plays a role during development of Parkinsonian symptoms, the course appears similar to that of an environmental insult. In this case the period of presymptomatic symptoms is short and the point of time with instrumental methods will be nearly the same as for clinical examination

Table 1. Approaches to early diagnosis and prediction of PD

— Neuroimaging
    PET
    SPECT
— Biochemical markers
— Genetic markers
— Instrumental methods
— Pharmacological tests
— Electrophysiology
— Risc factors

We have to differentiate between different phases of diagnosis: if there is not even an unspecific sign we talk about "prediction", if the diagnosis is ensured during the period of unspecific problems this procedure is called "early diagnosis" (Fig. 3).

Different methods can contribute to diagnosis during different periods: Genetic markers allow a real prediction e.g. in Huntington's disease, biochemical markers can be conspicuous from birth (e.g. phenylketonuria). Instrumental examinations of movement disorders operate in the phase of early diagnosis.

A biochemical or genetic marker would be most helpful if it were specific for the whole span of life and therefore would allow prediction or even a real screening. Unfortunately we don't have such result (for an overview about biochemical markers see e.g. Przuntek, 1989; Gerlach, 1992; Koller, 1991).

A different kind of imaginable markers are a result of the disease process and then measurable only after the process already is running.

Neuroimaging methods (PET, SPECT, MRI, CCT) can estimate the substantial loss of degeneration, but because of the large inter-individual variance

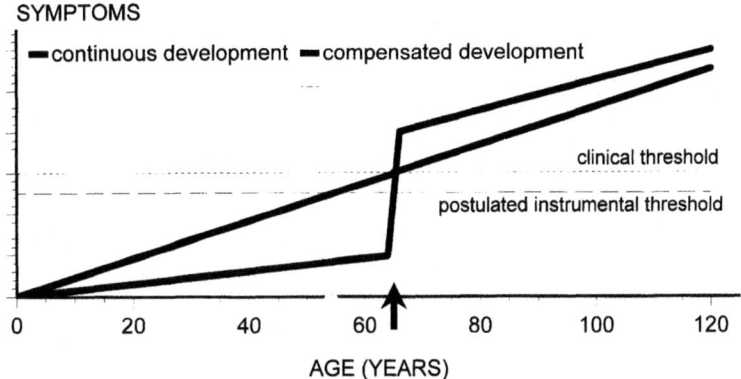

Fig. 3. Model of diagnosis in PD: Over the course of Parkinson's disease at different points of time we differentiate between prediction (even no unspecific symptoms), early diagnosis and clinical diagnosis. Different methods can contribute to diagnosis in different phases

of findings there are no normal values for early diagnosis. One parameter promising better selectivity than cross sectional comparison is the asymmetry of findings for left and right side (Seybyl, 1995).

There are first results of longitudinal PET examinations of patients with mild PD over about 2 years.

Leenders et al. found a decrease of about 9% per year for de novo PD patients and inferred the presymptomatic interval to be about 5 years (Leenders, 1995). But the different rate for decrease of 3% per year for patients suffering from PD for a longer time indicates an inconstant process. Following the results of Lee et al. (1991) the rate of neuronal death is more rapid in earlier stages in consistency with Leenders (1995). Morrish et al. (1995) found PET results which were compatible with a short preclinical period.

To improve selectivity of clinical and instrumental methods additional application of specific therapy may be helpful in early cases — improvement of symptoms after application of L-dopa is one of the essential features for diagnosis of PD. The so called "apomorphine test" which may be useful for differential diagnosis between PD and multi-system atrophies in later cases is not a sufficient tool during the phase of unspecific symptoms: Following the results of our examination of apomorphine in de novo patients with PD (with help of the motor performance test after Schoppe) showed a positive reaction which was unfortunately not specific compared with the improvement after placebo.

Until searching for a specific biochemical or genetic marker may be successful we can improve selectivity of established examinations by working out normal data by longitudinal examinations and looking for typical constellations (profiles) of parameters (which may be very different) with multivariate statistical methods.

References

Gerlach M, Riederer P (1992) Gibt es biochemische Marker der Parkinson-Krankheit? In: Fischer PA (Hrsg) Parkinson-Krankheit: verlaufsbezogene Diagnostik und Therapie. Editiones <Roche>, Basel/Grenzach-Wyhlen

Koller W, Langston JW (eds) (1991) Preclinical detection of Parkinson's disease. Neurology 41(5) [Suppl 2]

Lee CS, Schulzer M, Mak EK, Snow BJ, Tsui JK, Calne S, Hammerstad J (1994) Clinical observations on the rate of progression of idiopathic parkinsonism. Brain 117: 501–507

Leenders KL (1995) Oral presentation. Fifth meeting of the European Neurological Society

Morrish PK, Sawle GV, Brooks DJ (1995) Clinical and [18F]-DOPA PET findings in early Parkinson's disease. J Neurol Neurosurg Psychiatry 59(6): 597–600

Przuntek H, Riederer P (eds) (1989) Early diagnosis and preventive therapy in Parkinson's disease. Springer, Wien New York (Key Topics in Brain Research)

Segawa M, Nomura Y, Kase M (1986) Hereditary progressive Dystonia with marked diurnal fluctuation: clinicopathophysiological identification in reference to juvenile Parkinson's disease. Adv Neurol 45: 227–234

Seibyl JP, Marek KL, Quinlan D, Sheff K, Zoghbi S, Zea-Ponce Y, Baldwin RM, Fussell B, Smith EO, Charney DS, et al (1995) Decreased single-photon emission computed tomographic [123I]beta-CIT striatal uptake correlates with symptom severity in Parkinson's disease. Ann Neurol 38(4): 589–598

Author's address: Dr. P. H. Kraus, Department of Neurology, St. Josef Hospital, Gudrunstrasse 56, D-44791 Bochum, Federal Republic of Germany

J Neural Transm (1996) [Suppl] 48: 29–43
© Springer-Verlag 1996

The pharmacology of B-type selective monoamine oxidase inhibitors; milestones in (−)-deprenyl research

K. Magyar[1], **B. Szende**[2], **J. Lengyel**[3], and **K. Tekes**[1]

[1]Department of Pharmacodynamics, [2]1st Department of Pathology and Experimental Cancer Research, and [3]Central Isotope Laboratory, Semmelweis University of Medicine, Budapest, Hungary

Summary. (−)-deprenyl cannot be considered as a simple, selective inhibitor of MAO-B. It increases the dopaminergic tone in the central nervous system by a complex mechanism. The MAO-B inhibition could result in a potentiation of the effect and the reduction of the dose of L-dopa, including the restoration of the sensitivity to L-dopa treatment, when the response to the drug has already been diminished or lost. Pre-treatment with (−)-deprenyl prevent the effect of neurotoxins like MPTP, 6-hydroxydopamine, DSP-4, AF64A by inhibiting the conversion of the pretoxin to toxin, or by inhibiting the neuronal reuptake mechanisms, or the combination of the two processes. However, other effects of the inhibitor cannot be ruled out. (−)-deprenyl, but not its (+)-enantiomer, proved to be a potent inhibitor of programmed cell death (apoptosis) of PC12 cells and that of human melanoma cells, in a concentration which does not induce MAO-B inhibition. The activity of MAO-B increases with age and the age related changes led to an overproduction of neurotoxic agents. The inhibition of the enzyme activity can play a preventive role against neurodegenerative brain disorders. The most widely used MAO-B inhibitor in the therapy is (−)-deprenyl and it lacks the "cheese reaction". The complex mechanism for the lack of the former effect is not fully known.

Introduction

The discovery that inhibitors of monoamine oxidase (MAO; EC 1.4.3.4.) are potent antidepressants has resulted in the synthesis of large numbers of MAO inhibitors (for review: Tipton, 1989). It was recognised, that two types of MAO are existing in mammalian tissues, MAO-A and MAO-B, with different substrate and inhibitor specificities (Johnston, 1968). Inhibitors of MAO-A proved to be effective antidepressants, while MAO-B blockers appear to be useful in the treatment of Parkinson's disease. It has also been demonstrated that MAO-B inhibitors can either protect or rescue neurons from damage. MAO-A inhibitors when used as antidepressants can induce hypertensive response, following the ingestion of tyramin containing foods. Because of this potential harm A-type inhibitors have fallen into disrepute as antidepressants (for review Magyar, 1992).

The irreversible types of inhibitors are used most frequently in therapy, but the therapeutic application of the inhibitors of reversible type cannot be ruled out. The discovery and the therapeutic use of moclobemide can throw light upon the possible role of the A-type selective reversible inhibitors in the treatment of depression (Da Prada et al., 1990).

Deprenyl was first described in 1965 (Knoll et al.). Its (−)-enantiomer proved to be a potent irreversible inhibitor of MAO (Magyar et al., 1967), which inhibits specifically the B-type of the enzyme (Knoll and Magyar, 1972). The intensive studies with (−)-deprenyl revealed that it cannot be considered as a simple MAO-B inhibitor, in spite of the fact that in a very low concentration ($IC_{50} = 10^{-8}M$), it can potently inhibit MAO-B activity using phenylethylamine as a substrate.

Pharmacological activities of MAO-B inhibitors

Before giving a description of the pharmacological activities derived from MAO-B inhibition, it should be emphasised that some inhibitors have in addition to their enzyme inhibitory effect, further intrinsic pharmacological activities (like uptake inhibitory potency) not related to the inhibition of MAO-B.

Dopamine potentiating effect of MAO-B inhibitors

Although there is a considerable amount of confusion regarding the substrate specificity of MAO-A and MAO-B towards dopamine (DA), it is widely accepted that while DA is metabolised by both forms of the enzyme in the rat brain (Fowler and Strolin-Benedetti, 1983), it is deaminated solely by MAO-B in the human brain (Glover et al., 1977). This substrate specificity found in the human brain offered an opportunity using (−)-deprenyl as an adjunct to L-dopa therapy of Parkinson's disease (Birkmayer et al., 1975; Rinne, 1983). The inhibition of MAO-B by (−)-deprenyl impairs the degradation of DA and reduces the amount of L-dopa required to maintain the optimal DA level in the parkinsonian brain (Knoll, 1978; Reynolds et al., 1980). The L-dopa sparing effect of (−)-deprenyl seems to potentiate the antiakinetic effect of L-dopa and reduces the incidence of dyskinesias, as "on-off" phenomena (for review see; Da Prada et al., 1984).

The inhibition of MAO-B primarily occurs in the glia, which leads to a considerable increase of DA in the synaptic gap. It is quite likely that DA from the glia overflows in a hormone-like process and that it recognises its own synaptic receptors (Riederer et al., 1989). In addition to the selective inhibition of MAO-B, (−)-deprenyl seems to interfere with DA regulation by several other mechanisms, e.g. inhibiting the uptake of noradrenaline (NA) and DA (Knoll and Magyar, 1972).

In higher doses of (−)-deprenyl than those needed for selective inhibition of MAO-B, considerable amounts of amphetamine-like metabolites can be formed. Nevertheless, the releasing potencies of the (−)-isomers of the amphetamines produced by the metabolism from (−)-deprenyl are not equivalent to those of the (+)-enantiomers. This might partly be due to the fact that the release of NA from the depot granules is not the direct effect of the amphetamines, but rather that of their further metabolite, (+)-p-hydroxynorephedrine (Brodie et al., 1970). Since only the (+)-p-hydroxyamphetamine is converted to (+)-p-hydroxy-norephedrine in vivo [the (−)-p-hydroxyamphetamine is not a substrate for beta-hydroxylase] this can explain the difference between the releasing potencies of the amphetamine enantiomers (Goldstein and Anagnoste, 1965). Unlike the different releasing potencies of the stereoisomers, we found both amphetamine enantiomers practically equally effective in inhibiting the uptake process (Magyar and Knoll, 1970; Magyar, 1991).

The principal pharmacological influence of (−)-deprenyl on the dopaminergic neurones is the augmentation of their sensitivity to physiological and pharmacological stimuli, which effect is the consequence of the inhibition of both MAO-B activity and DA uptake (Knoll, 1983). After chronic treatment, (−)-deprenyl reduces the number of beta-adrenoceptors and increases the turnover rate of DA (Zsilla et al., 1983; Zsilla and Knoll, 1982). The MAO-B inhibition induced by (−)-deprenyl could result in a) the potentiation of L-dopa effect, b) the reduction of the daily dose of L-dopa, and c) restoration of the sensitivity to L-dopa treatment, when the response has already been diminished or lost (Youdim and Finberg, 1986).

Neuroprotective effect of MAO-B inhibitors

The neuroprotective effect of (−)-deprenyl has been demonstrated first by Birkmayer and his co-workers in a retrospective clinical trial (Birkmayer et al., 1985). They observed that parkinsonian patients treated with Madopar® plus (−)-deprenyl lived significantly longer than those taking Madopar® alone, due to a neuroprotective effect of the MAO-B inhibitor. After a long period of discussion of this finding, Tetrud and Langston (1989) proved recently in a well controlled prospective study on early parkinsonian patients, that (−)-deprenyl treatment delayed the need for levodopa therapy by slowing the disease's progression, as it was judged after one month "wash out" period.

The one month "wash out" phase was criticised in the DATATOP trial, in which 400 patients were treated with 10 mg of (−)-deprenyl per day and were compared in double-blind studies with matched controls on placebo. The need for levodopa therapy was delayed in the (−)-deprenyl treated groups, but it was argued that even after one month discontinuation of the drug the symptomatic effects of (−)-deprenyl were still present (Landau, 1990), whilst, a line of evidence suggests that the biochemical effects of (−)-deprenyl disappear rather rapidly after the cessation of the drug administration (Elsworth et al., 1978)

Neuroprotection induced by (−)-deprenyl against toxic insults

In recent years it has become also apparent that (−)-deprenyl can protect neurones from a variety of toxins, which induce neurodegeneration. The neuroprotective effect of (−)-deprenyl against MPTP (1-methyl-4-phenyl-1,2,3,6 tetrahydropyridine), 6-hydroxydopamine and DSP-4 [N-(2-chloroethyl)-N-ethyl-2-bromobenzylamine] has been demonstrated. Similar protection due to (−)-deprenyl pre-treatment was shown against a central cholinergic neurotoxin, AF64A (methyl-β-acetoxyethyl-2-chloroethylamine) (Ricci et al., 1992).

MPTP toxicity

The mechanism of MPTP toxicity was excellently reviewed by Glover et al. (1986). The substance is a preferential substrate for MAO-B (Salach et al., 1984), as its oxidation is highly sensitive to inhibition by (−)-deprenyl, but not by clorgyline. A daily oral dose of 10mg of (−)-deprenyl to patients is sufficient to inhibit MPTP oxidation e.g. its conversion to the toxic MPP^+ (1-methyl-4-phenylpiridine). The toxin, MPP^+ formed by MAO-B from the pretoxin (MPTP), is actively taken up by the dopaminergic nerve terminals via the dopamine (DA) re-uptake processes (Javitsch et al., 1985). Since the formation of the neurotoxin MPP^+ from the parent compound MPTP is dependent on the MAO-B activity, all the selective inhibitors of MAO-B can potentially prevent MPTP-induced neurodegeneration in vivo.

Both forms of MAO are considered to play a general protective role by preventing the central and peripheral nervous system from exogenous amines. The conversion of MPTP to the active MPP^+ results in an opposite effect, than that of the normal function.

A line of evidence exists that inhibitors of DA uptake like desipramine and mazindol are also capable to prevent MPTP-induced neurodegeneration. In the rodent's brain MAO-B activity is located mainly in glial cells, but not in the dopaminergic nerve terminals. It was suggested that MPTP is taken up into the glial cells, where it is oxidised to MPP^+. Than the toxin is released and taken up by the dopaminergic neurones by an active re-uptake process, which mechanism can be inhibited by mazindol (Javitsch et al., 1985).

Since (−)-deprenyl and mainly its metabolites are potent inhibitors of DA uptake, this effect of (−)-deprenyl may play a considerable role in prevention of MPTP-induced neurotoxicity in addition to its MAO-B inhibitory action (Hársing et al., 1979; Magyar, 1991).

MPTP-induced toxicity is still the best primate model, of parkinsonian syndrome. Since this compound does not exist in the enviroment, it cannot be considered to be the cause of the idiopathic disease. Intensive research has been carried out to find endogenously generated neurotoxins or pretoxins, which can be activated by MAO-B. Nagatsu and his co-workers (1987) suggested tetrahydro-isoquinoline (TIQ) to be a candidate to produce parkinsonism. TIQ is methylated in the human brain and the product is a substrate for MAO-B which converts it to N-methyl-isoquinoline (NMIQ$^+$), an analogue of MPTP (Naoi et al., 1989). However, the existence of endogenous toxins causing parkinsonian syndrome remains to be proved.

6-Hydroxydopamine toxicity

The 6-hydroxydopamine induced nigro-striatal degeneration can also be prevented by pre-treatment with (−)-deprenyl (Knoll, 1987). The mechanism underlying the neural degeneration depends on the formation of 6-hydroxyquinone from 6-hydroxydopamine, which step is followed by an uptake into the dopaminergic nerve endings. 6-hydroxyquinone initiates neural degeneration due to the generation of free radicals. It was suggested that the protective effect of (−)-deprenyl can be due to the combination of at least three factors: inhibition of MAO-B activity, potentiation of the free radical scavenging systems, and inhibition of DA re-uptake process (Knoll, 1987).

MAO-B activity

The role of *MAO-B activity* in the activation of 6-hydroxydopamine is rather uncertain; so the inhibition of the enzyme by (−)-deprenyl cannot play a significant role in the prevention of 6-hydroxydopamine neurotoxicity.

Superoxide dismutase

It was published recently that chronic (−)-deprenyl treatment may increase *superoxide dismutase* (SOD) activity (Knoll, 1988; Clow et al., 1991). This effect in a combination with an increased activity of either catalase or glutathione peroxidase would improve free radical removal (Berry et al., 1994). Carrillo et al. (1991), were also able to show an increase in both cytosolic and particulate forms of SOD, with an elevated activity of catalase. However, a recent study failed to demonstrate any increase in SOD activity (Lai et al., 1994).

Noradrenaline uptake

(−)-deprenyl is a weak inhibitor of *noradrenaline uptake*. Its IC$_{50}$ is 3.6 × 10^{-5} mol/L in the synaptosomal preparation of the rat hypothalamus. S(+)-deprenyl was slightly more potent in this respect (IC$_{50}$ = 6.3 × 10^{-6} mol/L;

Tekes et al., 1988). Similar results were published by Knoll and Magyar (1972) in mouse cerebral cortical slices. Nevertheless, the uptake inhibition by the metabolites of (−)-deprenyl (methylamphetamine and amphetamine) cannot be ruled out as a contribution to the protective effect (see later).

Thus of the three suggested mechanisms of the protection against 6-hydroxydopamin toxicity, the inhibition of re-uptake appears to be the most important process.

DSP-4 toxicity

It was published recently, that (−)-deprenyl, but not MDL 72974/A [(E)-4-fluoro-β-(fluoromethylene)benzenebutanamine], another potent selective MAO-B inhibitor, was capable to prevent the depletion of noradrenaline (NA) in the mouse hippocampus induced by DSP-4 [N-(2-chloroethyl)-N-ethyl-2-bromobenzylamine] (Finnegan et al., 1990). Since both (−)-deprenyl and MDL 72974/A produce a comparable degree of MAO-B inhibition it seems doubtful that MAO-B activity play any significant role in the neurotoxicity of the substance.

DSP-4 produces a long-lasting inhibition of ^{3}H-noradrenaline (^{3}H-NA) uptake into the central and peripheral noradrenergic neurones of rodents (Ross et al., 1973; Ross and Renyi, 1976). It is a beta-haloethylamine derivative of benzylamine which interact with the presynaptic components of the adrenergic synapse. DSP-4 undergoes a spontaneous cyclization in solution to form a positively charged aziridinium ion. It was proposed that the irreversible damage of the noradrenergic neurones is due to the effect of the ion (Ross et al., 1973). The charged molecule forms covalent bond with the electrophilic centres present on its site of action and exerts irreversible damage, a widespread depletion of NA in various noradrenergic axon terminals.

DSP-4 is not converted to a toxic compound by MAO-B, like MPTP. It was suggested that the blockade of DSP-4 uptake by (−)-deprenyl may explain its protective effect (Finnegan et al., 1990). This proposition is supported by the findings, that desipramine also protects the NA nerve terminal against DSP-4 toxicity (Zicher and Jaim-Etcheverry, 1980). In addition to (−)-deprenyl, pargyline was also able to prevent DSP-4 induced neurotoxicity (Hallman and Jonsson, 1984), however, neither clorgyline, a selective MAO-A inhibitor (Gibson, 1987), nor the reversible inhibitor Ro 19-6327 (Bertocci et al., 1988) was effective.

Recently, a series of potent and selective aliphatic N-propargylamine type of MAO-B inhibitors lacking the potency to inhibit the uptake of dopamine and NA (Yu et al., 1994) were synthetised. Among these inhibitors 2-HxMP [N-(2-hexyl)-N-methylpropargylamine] was found to be the most potent MAO-B blocker, without an appreciable potency to inhibit the NA uptake. However, it was capable to prevent DSP-4 induced neurotoxicity. Based on these findings the authors concluded that the preventive role of MAO-B inhibitors against DSP-4 induced toxicity cannot be explained merely by their uptake inhibitory action. According to their studies MDL 72974/A in a relatively high concentration (10.0 mg/kg, i.p.) was also effective to prevent DSP-4 induced NA depletion, while in lower doses (1 to 2 mg/kg, i.p.), the substance was ineffective. In Finnegan's studies MDL 72974/A was administered in a dose of 1.25 mg/kg intraperitoneally (Finnegan et al., 1990). MDL 72145 another MAO-B inhibitor, having fluoroallilamine structure, also protected the hippocampal damage induced by DSP-4, at a higher dose (50 mg/kg i.p.)(Bertocci et al., 1988).

The effective doses of the inhibitors were certainly much higher than those required to inhibit MAO-B activity. These findings seem to support the view that MAO-B inhibition is not directly related to the mechanism of the restoration of hippocampal NA concentration.

Similar results to Finnegan's findings were obtained also in our laboratory in rats, concerning the neuroprotective effect of (−)-deprenyl (Magyar, 1991, 1994). DSP-4 was administered in a dose of 50 mg/kg intraperitoneally and the rats were decapitated 7 days after DSP-4 treatment. The NA content of the dissected hippocampus was determined by HPLC-ECD technique. DSP-4 treatment dropped the NA content of the rat hippocampus by 78%, but (−)-deprenyl pre-treatment in a dose of 10 mg/kg (i.p.) 1 h before DSP-4 administration, arrested the NA depletion in the hippocampus. (−)-deprenyl was slightly effective when it was injected 24 h before DSP-4 administration. The results are shown in Table 1.

According to the results of our studies the inhibition of the carrier mediated uptake process of NA by (−)-deprenyl plays more important role in the prevention of DSP-4 induced neurotoxicity, than it was declared in the publication of Yu and his co-workers (1994).

We have studied the effect of the stereoisomers of deprenyl and its potential metabolite methylamphetamine (MA), on the uptake of ^{3}H-NA, ^{3}H-dopamine (^{3}H-DA) and ^{3}H-serotonin (^{3}H-5-HT) in rat brain synaptosomes (Snyder and Coyle, 1969). The results (Table 2) show that in a reasonable concentration neither deprenyl nor MA influence the uptake of ^{3}H-5-HT, but both compounds inhibit the uptake of NA and DA. The (+)-enantiomers are more effective than the (−)-forms in this respect. From (−)-deprenyl only

Table 1. The effect of (−)-deprenyl pre-treatment on the DSP-4 induced NA depletion in the rat hippocampus

Pre-treatment	Treatment	NA content ng/mg tissue ± S.D.	%
Saline	Saline	0.32 ± 0.06	100
(−)-deprenyl	Saline	0.40 ± 0.06	125
Saline	DSP-4	0.07 ± 0.05	22
(−)-deprenyl (1 h*)	DSP-4	0.29 ± 0.04	90
(−)-deprenyl (24 h*)	DSP-4	0.11 ± 0.04	34

*pre-treatment time in h.; DSP-4 50 mg/kg i.p.; (−)-deprenyl 10 mg/kg i.p. Animals were decapitated 7 days after DSP-4 administration

(−)-MA can be formed in the human body (Schachter et al., 1980). On the basis of the IC_{50} values presented in Table 2 it can be concluded that the (−)-isomers are more potent to inhibit the uptake of NA than that of DA. The results also show that the (−)-form of the metabolite (MA) is 14.5 times more potent to inhibit of NA uptake than the parent compound.

Our uptake studies led us to the conclusion that most probably (−)-MA, the metabolite of (−)-deprenyl, is responsible for the neuroprotective effect against DSP-4. The ineffectiveness of MDL 72974 can be due to the fact that in the concentration administered neither the MDL compound nor its possible metabolites possesses uptake inhibitory properties.

The uptake inhibitory action of (−)-deprenyl and of its metabolite is probably a reversible process in spite of the irreversible MAO-B inhibition. That can explain by pharmacokinetic properties why (−)-deprenyl was found to be ineffective against DSP-4 toxicity, when it was injected 24h before DSP-4 treatment, when MAO-B inhibition was still complete. It is most probable that the uptake inhibitory effect of (−)-deprenyl or of its metabolites can also be responsible for preventing the toxicity of the cholinergic toxin AF64A, although the authors did not reach this conclusion (Ricci et al., 1992).

Hallman and Johnsson reported earlier (1984) that (+)-amphetamine, an uptake blocker of noradrenaline, which in this respect is more potent than (−)-deprenyl, was unable to protect the hippocampus from DSP-4-induced NA depletion. In contrast to their findings we demonstrated that both stereoisomers of MA are able to prevent quite effectively the DSP-4 induced toxicity even in a dose of 1mg/kg, when rats were treated intraperitoneally with the methylamphetamines 1h before 50mg/kg of DSP-4 administration. It has to be added that combination of DSP-4 with methylamphetamines gives rise to a rather toxic interaction. When 5 to 10mg/kg of (−)- or (+)-methylamphetamine was given with 50mg/kg of DSP-4, all the rats died, but with 1mg/kg of MA-s a good deal of prevention was achieved (Table 3).

We intensively studied the fate of (−)-deprenyl in rats by using the positionally and alternatively labelled radioisomers of the inhibitor (Magyar, 1994). By analysing the plasma concentration-time curve of the two labels, an intensive first pass metabolism of the compound, forming amphetamine metabolites, was revealed after oral administration. It is worth also to mention that we found (−)-deprenyl more effective to prevent DSP-4 induced NA

Table 2. Inhibition of the uptake of NA, DA and 5-HT in the synaptosomal fraction prepared from subcortical tissues of rats, by the enantiomers of deprenyl and methylamphetamine

Compounds	^{3}H-NA Hypothalamus	IC_{50} in M ^{3}H-DA Corpus striatum	^{3}H-5-HT Hippocampus
(−)-deprenyl	5.1×10^{-5}	1.0×10^{-4}	5.0×10^{-3}
(+)-deprenyl	1.7×10^{-5}	2.4×10^{-5}	3.6×10^{-2}
(−)-MA	3.5×10^{-6}	4.2×10^{-5}	$>5 \times 10^{-2}$
(+)-MA	3.5×10^{-7}	6.0×10^{-7}	1.9×10^{-2}

Table 3. The effect of (+)- and (−)-methylamphetamine (MA) pre-treatment on the DSP-4 induced NA depletion in the rat hippocampus

No. of rats	MA 1 mg/kg i.p.	DSP-4 50 mg/kg i.p.	MAO-B inhibition % ± S.D.	NA content % ± S.D.
30	−	−	0	100
5	−	+	9.1 ± 1.6	10.0 ± 3.5
5	(+)-MA	+	23.6 ± 0.2	59.0 ± 5.0
5	(−)-MA	+	0.6 ± 0.5	61.0 ± 3.5

MA-s were administered 1 h before DSP-4 treatment

depletion, when it was administered orally and not intraperitoneally. In an oral dose of 0.5 to 1.0 mg/kg its protective effect was quite remarkable.

Neural rescue effect of (−)-deprenyl

In addition to the neuroprotective effect of (−)-deprenyl pre-treatment against toxins, it has become apparent that (−)-deprenyl administration following the toxic insults can rescue the damaged neurones. It has been found that administration of (−)-deprenyl (0.25 mg/kg) three times a week, starting three days after MPTP treatment, increased neuronal survival significantly (Tatton and Greenwood, 1991). This type of treatment schedule did not prevent activation of MPTP by MAO-B. It is probable that in the post-insult treatment (−)-deprenyl increases the trophic support of the damaged neurones, which is resulted in their longer survival time. It was also published that when rats were treated with 0.005 to 0.01 mg/kg of (−)-deprenyl after facial motoneuron axotomy, the treatment increased the motoneural survival without inhibition of brainstem MAO-B activity (Ansari et al., 1993). (−)-deprenyl was published to be capable to rescue hippocampal neurones in acute cerebral ischemia by unknown mechanisms (Barber et al., 1993).

It has also been shown that (−)-deprenyl increased the neuronal survival of PC12 cells in tissue culture. The withdrawal of serum and nerve growth factor induced apoptosis in PC12 cells, but (−)-deprenyl inhibited the programmed cell death in a concentration of less than 10^{-9} M. This rescue effect of (−)-deprenyl was independent of its MAO-B inhibition (Tatton et al., 1994). The (+)-enantiomer of deprenyl lacks this property.

In order to study the neuronal rescuing effect of (−)-deprenyl in our laboratory, we used M-1 human melanoma cells (Ladányi et al., 1990) which were plated in 6-well Grainer (Kremsmünster, Germany) plates containing glass cover slips. The culture medium was RPMI (GIBCO), supplemented with 10% fetal calf serum (Bioproduct, Gödöllö, Hungary). Cell number at plating was 1.5×10^5/well. The medium was changed 48 hours after plating and no more serum supplementation was given since that time. Five days after changing the medium, (−)-deprenyl was given to the cell cultures to reach the final concentrations of 10^{-3}, 10^{-7} and 10^{-13} M. Samples were taken 24, 48 and 72 hours after treatment. Duplicate cover slip cultures per dose and per day, including 2-2 untreated controls were stained with Hematoxylin and Eosin. The percentage of apoptosis was determined taking into consideration the morphological signs of apoptosis (Wyllie et al., 1986).

The ratio of apoptotic cells increased gradually in the untreated cell cultures from the 24th to 72nd hours of the treatment period (Table 4). At 72 hours practically only apoptotic cells and cell debris were found (Fig. 1). (−)-deprenyl treatment significantly decreased the number of apoptotic cells even in the lowest concentration (Table 4). At 72 hours the cultures appeared to be viable (Fig. 2). (+)-deprenyl, however, did not prevent the high incidence of apoptosis (Table 4).

As a conclusion, the M-1 human melanoma cell line derives from melanocytes, which are also of neuroectodermal origin, were studied. According to our findings, (−)-deprenyl reduces apoptosis of serum-deprived M-1 cells. The other optical isomer, (+)-deprenyl fails to exert such effect.

The age-dependent increase of MAO-B activity

The MAO-B-age relationship was widely studied and it appears that brain MAO-B activity increases with age in both humans and animals (for review see; Strolin-Benedetti and Dostert, 1989). The increase of MAO-B activity with age may be associated with glial cell proliferation which is accompanied by neuronal loss, since microglia contains mostly the B form of MAO.

The enzyme MAO catalyses the oxidative deamination of primary, secondary, and tertiary amines, including neuronal transmitters, and aldehydes, ammonia, and hydrogen peroxide are formed. All of them are considered to be cytotoxic when they are not neutralised. The low content of reduced glutathione and the increased amount of ferrous ion in the substantia nigra of parkinsonian patients, and the hydrogen peroxide formed by deamination of dopamine might lead to an increase of oxygen free radicals which can elicit lipid peroxidation, and cause membrane damage and cell degeneration (for review see; Da Prada, 1991).

Table 4. The effect of (−)-deprenyl and (+)-deprenyl on apoptosis of M-1 cell cultures (Apoptotic index %)

Hours after treatment	Control sample		(−)-deprenyl 10^{-3}M sample		10^{-7}M sample		10^{-13}M sample		(+)-deprenyl 10^{-3}M sample		10^{-7}M sample		10^{-13}M sample	
	1	2	1	2	1	2	1	2	1	2	1	2	1	2
24	25	28	5	7	5	4	6	8	22	26	25	27	24	26
48	62	69	9	11	9	8	10	12	58	65	66	68	62	64
72	98	95	14	16	12	15	16	18	88	94	95	97	92	98

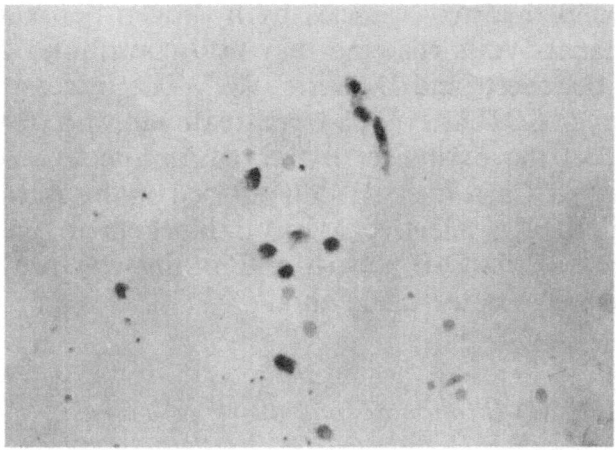

Fig. 1. Untreated M-1 human melanoma cell culture 10 days after plating and 8 days after serum deprivation (HE × 300)

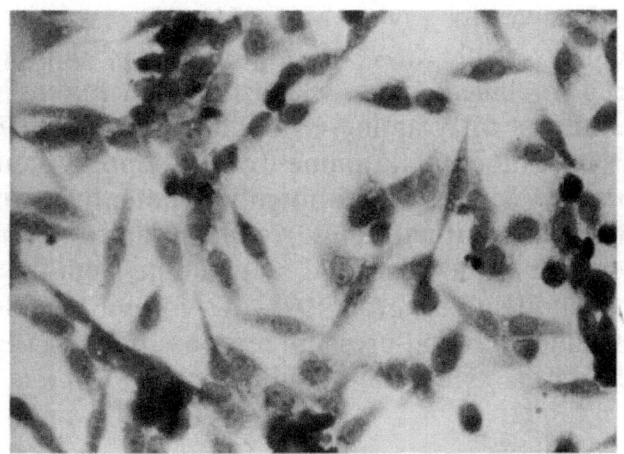

Fig. 2. 10^{-7} mol (−)-deprenyl-treated M-1 human melanoma cell culture 10 days after plating, 8 days after serum deprivation and 3 days after treatment (HE × 300)

There are conflicting data, but MAO-B activity in the blood platelets of parkinsonian patients and those with Alzheimer-type dementia (both of which are considered to be neurodegenerative diseases) is significantly higher than that in controls (Danielczyk et al., 1988; for review see Riederer et al., 1989).

MAO-B inhibitors, by reducing the formation of hydrogen peroxide from dopamine, might decrease the risk of neurodegeneration of the aging brain (Knoll, 1985; Langston, 1980). In addition to the oxidative stress caused by the age-dependent increase of MAO-B activity, further reactions, of endogenous amines and aldehydes produced by MAO-B can also play a role in neurodegeneration (Glover and Sandler, 1986).

The formation of hydroxy radicals by hydrogen peroxide through the Fenton or the Haber-Weiss reaction may also contribute to the neuronal damage (Strolin-Benedetti and Dostert, 1989). The increase in the radical scavenging activity of SOD after long-term treatment with (−)-deprenyl may additionally support the existence of the neuroprotective activity of (−)-deprenyl (Knoll, 1988; Carrillo et al., 1991). Based on the above observations, it seems rational to use selective MAO-B blockers to prevent the age-dependent increase of MAO-B activity and, in this way prevent the subsequent oxidative damage in the ageing brain.

MAO-B inhibitors and the "cheese effect"

The dominant opinion is that the non-selective irreversible MAO inhibitors and the MAO-A blockers are effective to treat psychic depression (Murphy et al., 1984), but their interactions with tyramine-containing foodstuffs may lead to serious blood-pressure response ("cheese effect"). MAO-B inhibitors generally lack the "cheese effect", but their effectiveness to treat depression is more than questionable (for review Magyar, 1992).

Among the irreversible MAO-B inhibitors (−)-deprenyl is the most widely used drug in clinical trials, which does not potentiate but strongly antagonises the effects of tyramine (Knoll and Magyar, 1972). The lack of potentiation of sympathomimetic amines by (−)-deprenyl could be attributed to several factors: (−)-deprenyl does not inhibit potently the intestinal MAO-A in man, it inhibits neuronal amine reuptake and does not release, but inhibit, noradrenaline outflow induced by tyramine from the storage vesicles of the rat heart (Knoll and Magyar, 1972). To form a reliable opinion about the lack of inducibility of blood-pressure response by MAO-B blockers, more clinical experience is needed with the inhibitors.

Acknowledgement

The experimental part of the paper was supported by the grant OTKA T 017749 of the Hungarian Academy of Sciences.

References

Ansari KS, Tatton WG, Yu PH, Kruck TPA (1993) Rescue of axotomised immature rat facial motoneurons by R(−)-deprenyl: stereospecificity and independence from monoamine oxidase inhibition. J Neurosci 13: 4042–4053

Barber AJ, Paterson IA, Gelowitz DL, Voll CL (1993) Deprenyl protects rat hippocampal pyramidal cells from ischaemic insult. Soc Neurosci Abstr 19: P1646

Berry MD, Juorio AV, Paterson IA (1994) The functional role of monoamine oxidases A and B in the mammalian central nervous system. Prog Neurobiol 42: 375–391

Bertocci B, Gill G, Da Prada M (1988) Prevention of the DSP-4-induced noradrenergic neurotoxicity by irreversible, not by reversible MAO-B inhibitors. Pharmacol Res Commun 20: 131–132

Birkmayer W, Riederer P, Youdim MBH, Linauer W (1975) The potentiation of the anti-akinetic effect of L-dopa treatment by an inhibitor of MAO-B, l-deprenyl. J Neural Transm 36: 303–336

Birkmayer W, Knoll J, Riederer P, Youdim MBH, Hars V, Martin J (1985) Increased life expectancy resulting from addition of l-deprenyl to Madopar® treatment in Parkinson's disease: a long-term study. J Neural Transm 64: 113–128

Brodie BB, Cho AK, Gessa GL (1970) Possible role of p-hydroxynorephedrine in the depletion of norepinephrine induced by d-amphetamine and in tolerance to this drug. In: Costa E, Garattini S (eds) Amphetamines and related compounds. Raven Press, New York, pp 217–230

Carrillo MC, Kanai S, Nokubo M, Kitani K (1991) (−)-Deprenyl induces activities of both superoxide dismutase and catalase but not of glutathione peroxidase in the striatum of young male rats. Life Sci 48: 517–521

Clow A, Hussain T, Glover V, Sandler M, Dexter DT, Walker M (1991) (−)-deprenyl can induce soluble superoxide dismutase in rat striata. J Neural Transm 86: 77–80

Da Prada M (1991) New approaches to the treatment of age-related brain disorders. Can J Neurol Sci 18: 384–386

Da Prada M, Keller HH, Pieri L, Kettler R, Haefely WE (1984) The pharmacology of Parkinson's disease: basic aspects and recent advances. Experiencia 40: 1165–1172

Da Prada M, Kettler R, Keller HH, Cesura AM, Richards JG, Marti JS (1990) From moclobemide to Ro 19-6327 and Ro 41-1049: the development of a new class of reversible, selective MAO-A and MAO-B inhibitors. J Neural Transm [Suppl] 29: 279–292

Danielczyk W, Streifler M, Konradi C, Riederer P, Moll G (1988) MAO-B activity and the psychopathology of Parkinson's disease, senile dementia and multiinfarction dementia. Acta Psychiatr Scand 78: 730–736

Elsworth JD, Glover V, Reynolds GP, Sandler M, Lees AJ, Phuapradit P, Shaw KM, Stern GM, Kumar P (1978) Deprenyl administration in man; a selective monoamine oxidase B inhibitor without the "Cheese effect". Psychopharmacology 57: 33–38

Finnegan KT, Skratt JS, Irwin I, DeLanney LE, Langston JW (1990) Protection against DSP-4-induced neurotoxicity by deprenyl is not related to its inhibition of MAO-B. Eur J Pharmacol 184: 119–126

Fowler CJ, Strolin-Benedetti M (1983) The metabolism of dopamine by both forms of monoamine oxidase in the rat brain and its inhibition by cimaxatone. J Neurochem 40: 1534–1541

Gibson CJ (1987) Inhibition of MAO-B, but not MAO-A blocks DSP-4 toxicity on central NE neurons. Eur J Pharmacol 141: 135–138

Glover V, Sandler M (1986) Clinical chemistry of monoamine oxidase. Cell Biochem Func 4: 89–97

Glover V, Sandler M, Owen F, Riley GJ (1977) Dopamine is a monoamine oxidase B substrate in man. Nature 265: 80–81

Glover V, Gibb C, Sandler M (1986) The role of MAO in MPTP toxicity — a review. J Neural Transm [Suppl] 65–76

Goldstein M, Anagnoste B (1965) The conversion in vivo of d-amphetamine to (+)-p-hydroxynorephedrine. Biochim Biophys Acta 107: 166–168

Hallman H, Jonsson G (1984) Pharmacological modifications of the neurotoxic action of the noradrenaline neurotoxin DSP-4 on central noradrenaline neurons. Eur J Pharmacol 103: 269–278

Harsing LG, Magyar K, Tekes K, Vizi ES, Knoll J (1979) Inhibition by deprenyl of dopamine uptake in rat striatum: a possible correlation between dopamine uptake and acethylcholine release inhibition. Pol J Pharmacol Pharm 31: 297–307

Javitch JA, d'Ámato RJ, Strittmatter SM, Snyder SH (1985) Parkinsonism-inducing neurotoxin, N-methyl-4-phenyl-1,2,3,6-tetrahydropyridine: uptake of the metabolite N-methyl-4-phenylpyridine by dopamine neurons explains selective toxicity. Proc Natl Acad Sci USA 82: 2173–2177

Johnston JP (1968) Some observations upon a new inhibitor of monoamine oxidase in brain tissue. Biochem Pharmacol 17: 1285–1297

Knoll J (1978) The pharmacology of selective irreversible monoamine oxidase inhibitors. In: Seiler N, Jung MJ, Koch-Weser J (eds) Enzyme activated irreversible inhibitors. Elsevier-North Holland, Amsterdam, pp 253–269

Knoll J (1983) (−)-Deprenyl ((−)-deprenyl): the history of its development and pharmacological action. Acta Neurol Scand [Suppl] 95: 57–80

Knoll J (1985) Striatal dopamine, aging and deprenyl. In: Borsy J, Kerecsen L, György L (eds) Proc 4th Cong Hung Pharmacol Soc Budapest 3: 7–25

Knoll J (1987) R-(−)-Deprenyl ((−)-deprenyl, Mogervan®) facilitates the activity of the nigro-striatal dopaminergic neuron. J Neural Transm 25: 45–66

Knoll J (1988) The striatal dopamine dependency of life span in male rats. Longevity study with (−)-deprenyl. Mech Aging Dev 46: 237–262

Knoll J, Magyar K (1972) Some puzzling pharmacological effects of monoamine oxidase inhibitors. Adv Biochem Psychopharmacol 5: 393–408

Knoll J, Ecseri Z, Kelemen K, Nievel J, Knoll B (1965) Phenylisopropylmethylpropinylamine (E-250), and new spectrum psychic energizer. Arch Int Pharmacodyn Ther 155: 154–164

Ladányi A, Timár J, Paku S, Molnár G, Lapis K (1990) Selection and characterization of human melanoma lines with different liver-colonizing capacity. Int J Cancer 46: 456–461

Lai CT, Zuo DM, Yu PH (1994) Is brain superoxide dismutase activity increased following chronic treatment with L-deprenyl? J Neural Transm 41: 221–229

Landau WM (1990) Clinical neuromythology IX. Pyramid sale in the bucket shop: DATATOP bottoms out. Neurology 40: 1337–1339

Langston JW (1980) (−)-deprenyl as neuroprotective therapy in Parkinson's disease: concepts and controversies. Neurology [Suppl 3] 40: 61–66

Magyar K (1991) Neuroprotective effect of deprenyl and p-fluor-deprenyl. Second Congress Paneuropean Society of Neurology, Vienna, p26

Magyar K (1992) Pharmacology of monoamine oxidase type B inhibitors. In: Szelenyi I (ed) Inhibitors of monoamine oxidase B. Birkhäuser, Basel, pp 125–143

Magyar K (1994) Behaviour of (−)-deprenyl and its analogues. J Neural Transm 41: 167–175

Magyar K, Knoll J (1970) Effect of phenyl-isopropyl-methyl-propinylamine (deprenaline) on the subcellular distribution of ^{3}H-noradrenaline. Acta Physiol Hung 37: p414

Magyar K, Vizi ES, Ecseri Z, Knoll J (1967) Comparative pharmacological analysis of the optical isomers of phenyl-isopropyl-methyl-propinylamine (E-250). Acta Physiol Hung 32: 377–387

Murphy DL, Sunderland T, Cohen RM (1984) Monoamine oxidase-inhibiting antidepressants. Psychiat Clin North Am 7: 549–562

Nagatsu T, Hirata Y (1987) Inhibition of the tyrosine hydroxylase system by MPTP, 1-methyl-4-phenylpiridinium ion (MPP$^+$) and the structurally related compounds in vitro and in vivo. Eur Neurol 26 [Suppl 1]: 11

Naoi M, Matsuura S, Takahashi T, Nagatsu T (1989) An N-methyltransferase in human brain catalyses N-methylation of 1,2,3,4-tetrahydroisoquinoline into N-methyl-1,2,3,4-tetrahydroisoquinoline, a precursor of a dopaminergic neurotoxin, N-methylisoquinolinium ion. Biochem Biophys Res Commun 161: 1213–1219

Reynolds GP, Riederer P, Rausch WD (1980) Dopamine metabolism in human brain: effects of monoamine oxidase inhibition in vitro by (−)-deprenyl and (−)-tranylcypromine. J Neural Transm [Suppl] 16: 173–178

Ricci A, Mancini M, Strocchi P, Bongrani S, Bronzetti E (1992) Deficits in cholinergic neurotransmission markers induced by ethylcholine mustard aziridinium (AF64A) in the rat hippocampus: sensitivity to treatment with the monoamine oxidase-B inhibitor l-deprenyl. Drugs Exp Clin Res VIII(5): 163–171

Riederer P, Konradi C, Hebenstreit G, Youdim MBH (1989) Neurochemical perspectives to the function of monoamine oxidase. Acta Neurol Scand 126: 41–45

Rinne UK (1983) A new approach to the treatment of Parkinson's disease. Acta Neurol Scand [Suppl 95] 68: 5–144

Ross SB, Renyi AL (1976) On the long-lasting inhibitory effect of N-(2-chloroethyl)-N-ethyl-2-bromobenzylamine (DSP-4) on the active uptake of adrenaline. J Pharm Pharmacol 28: 458–459

Ross SB, Johansson JG, Lindborg B, Dahlbom R (1973) Cyclizing compounds. I. Tertiary N-(2-chloroethyl)-N-ethyl-2-haloalkylamine with adrenergic blocking actions. Acta Pharm Suec 10: 29–42

Salach JI, Singer TP, Castagnoli N, Trevor A (1984) Oxidation of the neurotoxic amine 1-methyl-4-phenyl-1,2,3,6-tetrahydropyridine (MPTP) by monoamine oxidases A and B and suicide inactivation of the enzymes by MPTP. Biochem Biophys Res Comm 125: 831–835

Schachter M, Marsden CD, Parkes JD, Jenner P, Testa B (1980) Deprenyl in the management of response fluctuations in patients with Parkinson's disease on levodopa. J Neurol Neurosurg Psychiatry 43: 1016–1021

Snyder SR, Coyle JT (1969) Regional differences in ^{3}H-norepinephrine and ^{3}H-dopamine uptake into rat brain homogenates. J Pharmacol Exp Ther 165: 78–86

Strolin-Benedetti M, Dostert P (1989) Monoamine oxidase, brain ageing and degenerative diseases. Biochem Pharmacol 38: 555–561

Tatton WG, Greenwood CE (1991) Rescue of dying neurons: a new action for deprenyl in MPTP parkinsonism. J Neurosci Res 30: 666–672

Tatton WG, Ju WYL, Holland DP, Tai C, Kwan M (1994) (−)-Deprenyl reduces PC12 cell apoptosis by inducing new protein synthesis. J Neurochem 63: 1572–1575

Tekes K, Tóthfalusi L, Gaál J, Magyar K (1988) Effect of MAO inhibitors on the uptake and metabolism of dopamine in rat and human brain. Pol J Pharmacol Pharm 40: 653–658

Tetrud JW, Langston JW (1989) The effect of deprenyl ((−)-deprenyl) on the natural history of Parkinson's disease. Science 245: 519–522

Tipton KF (1989) In: Tipton KF, Youdim MBH (eds) Biochemical and pharmacological aspects of depression. Taylor & Francis, London, pp 1–24

Wyllie AH, Kerr JFR, Currie AR (1986) Cell death: the significance of apoptosis. Int Rev Cytol 68: 251–306

Youdim MBH, Finberg JPM (1986) MAO type B inhibitors as adjunct to L-dopa therapy. Adv Neurol 45: 127–136

Yu PH, Davis BA, Fang J, Boulton AA (1994) Neuroprotective effects of some monoamine oxidase-B inhibitors against DSP-4 induced noradrenaline depletion in the mouse hippocampus. J Neurochem 63: 1820–1828

Zieher LM, Jaim-Etcheverry G (1980) Neurotoxicity of N-(2-chloroethyl)-N-ethyl-2-bromobenzylamine (DSP-4) on noradrenergic neurons is mimicked by its cyclic aziridinium derivative. Eur J Pharmacol 65: 249–256

Zsilla G, Knoll J (1982) The action of (−)-deprenyl on monoamine turnover rate in rat brain. Adv Biochem Psychopharmacol 31: 211–217

Zsilla G, Barbaccia ML, Gandolfi O, Knoll J, Costa E (1983) (−)-Deprenyl, a selective MAO-B inhibitor increased ^{3}H-imipramine binding and decreased beta-adrenergic receptor function. Eur J Pharmacol 11: 117

Authors' address: Dr. K. Magyar, Department of Pharmacodynamics, Semmelweis University of Medicine, Üllöi U. 26, H-1085 Budapest VIII, Hungary

J Neural Transm (1996) [Suppl] 48: 45–59

(−)-Deprenyl reduces neuronal apoptosis and facilitates neuronal outgrowth by altering protein synthesis without inhibiting monoamine oxidase

W. G. Tatton[1,2,4]**, J. S. Wadia**[5]**, W. Y. H. Ju**[1]**, R. M. E. Chalmers-Redman**[1,4]**, and **N. A. Tatton**[3,4]

Departments of [1]Physiology/Biophysics, [2]Psychology, and [3]Anatomy/Neurobiology and [4]Institute for Neuroscience, Dalhousie University, Halifax, Nova Scotia, and [5]Department of Physiology, University of Toronto, Toronto, Ontario, Canada

Summary. (−)-Deprenyl stereospecifically reduces neuronal death even after neurons have sustained seemingly lethal damage at concentrations too small to cause monoamine oxidase-B (MAO-B) inhibition. (−)-Deprenyl can also influence the process growth of some glial and neuronal populations and can reduce the concentrations of oxidative radicals in damaged cells at concentrations too small to inhibit MAO. In accord with the earlier work of others, we showed that (−)-deprenyl alters the expression of a number mRNAs or proteins in nerve and glial cells and that the alterations in gene expression/protein synthesis are the result of a selective action on transcription. The alterations in gene expression/protein synthesis are accompanied by a decrease in DNA fragmentation characteristic of apoptosis and the death of responsive cells. The onco-proteins Bcl-2 and Bax and the scavenger proteins Cu/Zn superoxide dismutase (SOD1) and Mn superoxide dismutase (SOD2) are among the 40–50 proteins whose synthesis is altered by (−)-deprenyl. Since mitochondrial ATP production depends on mitochondrial membrane potential (MMP) and mitochondrial failure has been shown to be one of the earliest events in apoptosis, we used confocal laser imaging techniques in living cells to show that the transcriptional changes induced by (−)-deprenyl are accompanied by a maintenance of mitochondrial membrane potential, a decrease in intramitochondrial calcium and a decrease in cytoplasmic oxidative radical levels. We therefore propose that (−)-deprenyl acts on gene expression to maintain mitochondrial function and to decrease cytoplasmic oxidative radical levels and thereby to reduce apoptosis. An understanding of the molecular steps by which (−)-deprenyl selectively alters transcription may contribute to the development of new therapies for neurodegenerative diseases.

Introduction

Apparent clinical benefits of (−)-deprenyl in neurodegenerative diseases

Based on the capacity of (−)-deprenyl to inhibit monoamine oxidase B, (−)-deprenyl was first used in Parkinson's Disease (PD) as an adjunct to levodopa therapy (Birkmayer et al., 1977; Birkmayer and Riederer, 1984; Brannan and Yahr, 1995) in order to reduce the dose of levodopa and to decrease response fluctuations. Since 1989, a number of clinical trials have shown that (−)-deprenyl monotherapy can induce an apparent slowing of the progression of disability in the early stages of PD [see (Tetrud and Langston, 1989; Parkinson, 1993) for examples]. (−)-Deprenyl may also slow cognitive decline in mild to moderate Alzheimers Disease (AD) (see (Tariot et al., 1987; Mangoni et al., 1991; Martignoni et al., 1991) for examples of trials showing benefits of (−)-deprenyl in AD).

The capacity of (−)-deprenyl to slow the progression of PD and AD, although statistically significant, was at best moderate and in PD appeared to wane somewhat after one year of treatment. The design of the clinical trials did not provide for a determination the cellular or molecular mechanisms underlying the apparent slowing in the neurodegenerative diseases. Either one of two hypothetical mechanisms have been favored by the scientific community: 1) that increased dopaminergic neurotransmission could occur by inhibiting the oxidative deamination of dopamine to 3,4-dihydroxyphenylacetic acid (DOPAC) by MAO-B and thereby increasing the available pool of dopamine in nerve terminals (Birkmayer et al., 1977; Glover et al., 1977), and 2) that (−)-deprenyl might protect neurons from oxidative damage and death by reducing the production of H_2O_2 by inhibiting MAO-B catalyzed metabolism of dopamine (Cohen and Spina, 1989; Olanow, 1990). Two other possible MAO-B dependent mechanisms, although less prominently considered in the literature proposed: 1) that since (−)-deprenyl treatment increases striatal phenylethyamine by reducing its degradation by MAO-B, the increased phenylethlamine would facilitate activation of striatal dopamine receptors by dopamine (Paterson et al., 1990), and 2) that since (−)-deprenyl had been shown by some workers to increase the striatal activity of the scavenger enzymes like Cu/Zn superoxide dismutase (SOD1), Mn superoxide dismutase (SOD2), or catalase (Carrillo et al., 1990; Knoll, 1992) in some experimental animals, the increased scavenger activity would protect neurons from oxidative radical damage and death.

Calne and coworkers have argued the delay required for (−)-deprenyl to begin to slow the progression of PD and the gradual weakening of its action as judged from Kaplan-Meier plots implicate symptomatic mechanisms like a decrease in dopamine metabolism by MAO-B inhibition or improved dopaminergic neurotransmission by phenylethylamine as the likely mechanisms and rule against a decrease in oxidative radical damage (Schulzer et al., 1992). Alternately, a recent multiple stage clinical study by Olanow and coworkers (Olanow et al., 1995) seems to favor a neuroprotective action mediated by mechanisms like decreased oxidative radical production or increased scavenging by agents like SOD1.

What is the mehanism(s) by which (−)-deprenyl slows neurodegeneration?

Johnston (1968) first recognized two forms of brain monoamine oxidase, MAO-A and MAO-B and cloning of genomic DNAs for the two enzymes have shown differences in their promotor regions and in portions of their exons establishing that they are encoded at separate gene loci, differ in their amino acid sequences and can be separately controlled at a pre-translational level. (Bach et al., 1988; Grimsby et al., 1991; Shih, 1991). Both forms are localized to the outer leaflets of mitochondria with a smaller portion likely contained in the cytosol of neurons or glia (Mayanil and Baquer, 1984). The cellular localization of the two forms of MAO are likely important in determining the consequences of MAO-A versus MAO-B inhibition. Immunocytochemistry for MAO-A or MAO-B and histochemical techniques using reactions that specifically identify only one of the two forms have shown that MAO-A is present in most neurons while MAO-B is only found in serotonergic and some histaminergic neurons. MAO-B is also concentrated in non-neuronal cells, particularly astrocytes. Virtually identical cellular localizations have been found for the rodent (Demarest et al., 1980; Vincent, 1989), monkey (Westlund et al., 1985) and human brain (Westlund et al., 1988; Konradi et al., 1989). With relevance to PD, dopaminergic nigrostriatal neurons and other catecholaminergic neurons are negative for MAO-B and contain only MAO-A in all three species.

If MAO-B is not localized to human dopaminergic neurons and is only present in striatal astroglia, then extraneuronal or MAO-B independent mechanisms are needed to support the view that (−)-deprenyl increases striatal dopamine levels in PD. One possibility depends on the finding that (−)-deprenyl can inhibit the reuptake of catecholamines into terminals independently of MAO inhibition (Knoll and Mngjar, 1972) and therefore might increase extracellular dopamine levels in PD (Knoll, 1992). Inhibition of dopamine reuptake requires (−)-deprenyl concentrations unlikely to be reached with the doses of deprenyl used in PD (Gerlach et al., 1992) and therefore is unlikely to underlie the apparent slowing of the progression of PD. Furthermore, microdialysis of the rodent striatum has not shown acute changes in extracellular dopamine after (−)-deprenyl treatment while MAO-A inhibitors induce marked increases in extracellular dopamine (Butcher et al., 1990), which seems to argue against inhibition of dopamine reuptake or increased dopamine release due to decreased dopamine catabolism as the basis for neurodegenerative slowing. Whole tissue striatal dopamine and DOPAC levels, which reflect concentrations in both the extracellular and intracellular compartments, or dopamine release from striatal slices after (−)-deprenyl treatment in rats show no change after acute administration but do show increased dopamine content and decreased DOPAC release after chronic 3 week treatment (Dluzen and McDermott, 1991; Berry et al., 1994). Prolonged treatment with irreversible MAO-B inhibitors, like deprenyl, induces MAO-A inhibition, even at doses insufficient to acutely inhibit MAO-A (Waldmeier and Felner, 1978). Hence MAO-A rather than MAO-B inhibition could cause increased dopamine levels with prolonged (−)-deprenyl treatment and might explain the apparent slowing found in clinical trials.

(−)-Deprenyl is mainly metabolized to (−)-desmethyldeprenyl, (−)-methamphetamine and (−)-amphetamine (Karoum et al., 1982). The (−)-enantiomers of amphetamine and methamphetamine have a 10 times weaker effect on attention than (+)-methamphetamine and (+)-amphetamine (Taylor and Snyder, 1974) so that concentrations of the (−)-enantiomers produced by a 10 mg daily dose of (−)-deprenyl are considered too small to influence attention which could in turn account for improved cognitive performance in AD (Gerlach et al., 1992).

The validity of the proposal that (−)-deprenyl slows the progression of PD and AD by reducing oxidative radical damage depends on two factors: 1) that damage caused by oxidative radicals underlies the neuronal death in the two neurodegenerative diseases (see Youdim et al., 1994), either from increased free radical production or a failure of the scavenging mechanisms, 2) that inhibition of MAO-B by (−)-deprenyl does in fact decrease the production of oxidative radicals or increases radical scavenging. Dopaminergic neurons have been considered as particularly vulnerable to damage by oxidative radicals (Spina and Cohen, 1989). The catabolism of dopamine to DOPAC produces H_2O_2 and NH_3 while the autoxidation of dopamine to semi-quinones produces superoxide radicals (O_2^-) and H_2O_2. In the presence of Fe^{++}, H_2O_2 can be converted to hydroxyl radicals ($\cdot OH$) and hydroxyl ions (OH^-) (Halliwell, 1992). Neuronal loss in PD has been proposed to result in accelerated dopamine turnover which could be further accelerated by levodopa treatment (Carlsson and Fornstedt, 1991). Also since Fe^{++} accumulates in the substantia nigra in PD (Sofic et al., 1991; Good et al., 1992; Jellinger et al., 1992), possibly by an association with neuromelanin (Youdim et al., 1994), H_2O_2 produced from dopamine metabolism or dopamine autoxidation could be converted to hydroxyl radicals. Superoxide radicals and hydroxyl radicals produced by the above reactions might participate in lipid peroxidation with resultant neurodegeneration due to altered iron metabolism (Gerlach et al., 1994). Injection of $FeCl_3$ into the rat substantia nigra causes the death of dopaminergic neurons (Sengstock et al., 1992) as well as biochemical and behavioral changes typical of parkinsonism (Ben-Shachar and Youdim, 1991), presumably through the increased production of oxidative radicals. Some cortical areas of patients with AD have been found to contain increased levels of iron and ferritin (Dedman et al., 1992) which also might facilitate oxidative damage in that disease (Richardson et al., 1992).

Review and discussion of recent results

If MAO-B is not a neuronal enzyme (see above), MAO-B inhibition would not reduce the catabolism of dopamine to DOPAC within dopaminergic neurons in PD. Reductions in H_2O_2, O_2^- and $\cdot OH$ produced by MAO-B inhibition would occur in astrocytes. Protection from damage would then have to result from the decreased levels of toxic radicals escaping from the astrocytes in sufficient concentrations to produce lipid peroxidation of nearby neuronal membranes. Chiueh and coworkers have used microdialysis to show that (−)-deprenyl decreases extracellular hydroxyl radicals in the MPP+ damaged rat striatum (Wu et al., 1993; Chiueh et al., 1994). The decrease in radicals was induced with (−)-deprenyl concentrations as low as 10^{-11} M, too small to inhibit MAO-B and showing that (−)-deprenyl can reduce oxidative radical levels by a mechanism other than MAO-B inhibition.

Those findings were ascribed to a scavenger action of (−)-deprenyl (Wu et al., 1993), although it is difficult to envisage how such small (−)-deprenyl concentrations could significantly reduce radical concentrations without the aid of another scavenger. An increase in striatal superoxide dismutase (SOD1) was reported in rats treated with (−)-deprenyl (Knoll, 1988). Subsequent studies have found that the SOD1, SOD2 and catalase activity increase after 1–3 weeks of (−)-deprenyl treatment (Carrillo et al., 1990, 1992). The activity increases were regionally specific with parabolic dose relationships and indicated that the activity increases were independent of MAO-B inhibition (Carrillo et al., 1994). Similarly, (−)-deprenyl has been found to induce an acute increase in SOD1 and SOD2 activity in the MPTP-damaged mouse striatum (Thiffault et al., 1994). Most recently, dopaminergic indices have been used in an MPP+ striatal infusion model in the rat to show that (−)-deprenyl at concentrations and doses too small to inhibit MAO-B can "rescue" nigrostriatal neurons (Wu et al., 1995). Since the research indicates an action other than a decrease in oxidative radical production and the animals were examined at 4 days after MPP+ infusion allowing sufficient time for new protein synthesis, it is possible that (−)-deprenyl induced increases in radical scavengers, like SOD1, which are responsible for reducing the nigrostriatal damage.

(−)-Deprenyl increases neuronal survival in vivo and in vitro
without inhibiting MAO-B

Initially, it was thought that (−)-deprenyl only reduced the death of nigrostriatal dopaminergic neurons by inhibiting MAO-B and thereby blocking the conversion of MPTP to its active radical MPP+ (Heikkila et al., 1984; Langston et al., 1984). Subsequently, Mytilineou and Cohen (1985) showed that (−)-deprenyl maintained dopamine levels and dopamine uptake in mesencephalic explants after MPP+ treatment thereby establishing that (−)-deprenyl could reduce dopaminergic neuronal damage by a mechanism other than the blockade of the conversion of MPTP to MPP+. (−)-Deprenyl was shown to reduce the death of murine dopaminergic nigrostriatal neurons even when first administered 72 hour after MPTP treatment when MPTP was fully converted to MPP+ and maximal striatal damage had occurred (Tatton and Greenwood, 1991). It was then shown using the same delayed administration that (−)-deprenyl doses of 0.01 mg/kg every 2 days induced a maximal increase in the survival of the dopaminergic nigrostriatal neurons (Tatton et al., 1993). The 0.01 mg/kg dose was insufficient to inhibit MAO-A or MAO-B, even after 20 days of administration. It has also been shown to increase the survival of dopaminergic mesencephalic neurons in culture, either after trophic withdrawal (Roy and Bedard, 1993) or MPP+ treatment (Koutsiliere et al., 1994). (−)-Deprenyl was also shown to reduce DSP-4 toxicity which does not involve MAO-A or MAO-B (Finnegan et al., 1990; Yu et al., 1994; Zhang et al., 1995).

(−)-Deprenyl has been shown to reduce neuronal death in a variety of non-catecholaminergic neuronal types and after a variety of insults including: immature facial neurons after axotomy (Salo and Tatton, 1992; Ansari et al., 1993; Ju et al., 1994; Zhange et al., 1995), adult murine facial motoneurons after axotomy (Oh et al., 1994), rat retinal ganglion cells after optic nerve crush (Buys et al., 1995), rat CA1 hippocampal neurons after ischemia/hypoxia (Barber et al., 1993) and rat cerebellar Purkinje cells and granule cells with aging (Amenta et al., 1994). Studies with axotomized immature rat facial motoneurons showed that the maximum effectiveness of intraperitoneal (−)-deprenyl for increasing the survival of the motoneurons was greater than that with intrathecal delivery of ciliary neurotrophic factor (CNTF) and it did not induce the cachexia caused by CNTF (Henderson et al., 1994; Zhang et al., 1995). They also revealed that doses of (−)-deprenyl insufficient to inhibit MAO-B or MAO-A induced maximum increases in the survival of the immature facial motoneuron while doses of (+)-deprenyl sufficient to cause 80% MAO-B inhibition did not increase motoneuron survival (Ansari et al., 1993). With the exception of (−)-deprenyl and pargyline, most MAO-A or MAO-B inhibitors including iproniazid, phenelizine, semicarbazide, tranylcypramine, nialamide, MDL 72974A, RO-16-6491, clorgyline, and brofaromine did not increase the survival of the axotomized immature motoneurons (Tatton et al., 1992; Ansari et al., 1993). The ED_{50} for pargyline was shifted to about 3,000 fold higher doses than that of (−)-deprenyl (0.005 mg/kg every 2 days) and pargyline was significantly less effective in reducing motoneuron death (Zhang, Ansari and Tatton, unpublished observations). All in all, these results showed that the neuronal "rescue" action of (−)-deprenyl did not involve MAO-A or MAO-B inhibition and was dependent on a structural entity shared by (−)-deprenyl and pargyline but not (+)-deprenyl or other structurally similar compounds like clorgyline.

(−)-Deprenyl alters gene expression or protein synthesis for a variety of neuronal and glial genes/proteins

(−)-Deprenyl has been shown to alter glial fibrillary acidic protein (GFAP) levels in rat striatum and facial nucleus (Biagini et al., 1993, 1994; Ju et al., 1994) and GFAP mRNA in C6 glioma cells, CNTF mRNA in process bearing astrocytes in wounded cultures (Seniuk et al., 1994), basic fibroblast growth factor protein in activated astrocytes in vivo (Biagini et al., 1994), neurotrophin receptor (trk C) mRNA in the rat forebrain (Ekblom et al., 1993), superoxide dismutase activity (SOD1, SOD2) (Carrillo et al., 1990) in the rat striatum and hippocampus and aromatic amino acid decarboxylase (AAAD) in NGF supported PC12 cells (Li et al., 1992). The changes in protein synthesis/gene expression did not appear to follow a consistent pattern and the research did not allow for the determination as to whether they were directly induced by (−)-deprenyl or were secondary to another event caused by deprenyl like an alteration in oxidative radical levels or a change in monoamine levels.

(−)-Deprenyl increases process or neurite length in some glial cells and neurons in culture

(−)-Deprenyl increased the process length of explanted rat motoneurons even with concentrations as low as 10^{-12}M which are far too small to inhibit MAO-A or MAO-B (Iwasaki et al., 1994) and also maintained the neurite length of embryonal dopaminergic mesencephalic neurons after exposure to MPP+ (Koutsiliere et al., 1994). Similarly, (−)-deprenyl, but not (+)-deprenyl, increased the process length of process-bearing astrocytes at concentrations as low as 10^{-11}M.

The neuronal death in most of the models in which (−)-deprenyl has been shown to increase survival are now known to involve apoptosis (Wilcox et al., 1993; Berkelaar et al., 1994; Mochizuki et al., 1994; Rosenbaum et al., 1994). We therefore used a well characterized in vitro model of neuronal apoptosis, the apoptotic death of PC12 cells caused by trophic withdrawal (Rukenstein et al., 1991), to examine the capacity of (−)-deprenyl, (+)-deprenyl, (−)-deprenyl metabolites, and other MAO-B or -A inhibitors to reduce neuronal apoptosis (Tatton et al., 1994). The PC12 cells died gradually over five days after trophic withdrawal with about 50% dying in the first 24 hours. In situ marking of cut DNA 3′ ends using APOP TAG (Oncor Ltd.) and DNA electrophoresis revealed that the internucleosomal DNA fragmentation characteristic of apoptosis began in most cells by 8 to 12 hours after trophic withdrawal. (−)-Deprenyl markedly reduced both PC12 cell death and internucleosomal DNA fragmentation at concentration of 10^{-5} to 10^{-11}M according to a relationship with a maximum increase in survival at 10^{-9}M. Similar to our findings in axotomized facial motoneurons (Ansari et al., 1993), (+)-deprenyl did not alter PC12 cell survival showing that (−)-deprenyl interacted with a stereospecific site, other than the FAD site of MAO-B, to increase survival. MAO-A and MAO-B inhibitors like iproniazid, phenelzine, semicarbazide, tranylcypramine, nialamide, MDL 72974A, RO-16-6491, clorgyline, and brofaromine did not increase PC12 cell survival. Pargyline increased the survival with an IC_{50} about 1,000 fold greater than (−)-deprenyl. Two of the major metabolites of (−)-deprenyl, (−)-methamphetamine and (−)-amphetamine, increased the PC12 cell apoptosis but only at very high concentrations (10^{-3} to 10^{-5}M). More importantly, both (−)-methamphetamine and (−)-amphetamine antagonized the capacity of (−)-deprenyl to increase the survival of trophically-deprived PC12 cells in a dose dependent fashion (Tatton and Holland, unpublished observations). Parallel studies in immature axotomized facial motoneurons also revealed the antagonism of (−)-methamphetamine and (−)-amphetamine (Ansari, Zhang, Ju and Tatton, unpublished observations). Three different general blockers of P450 enzymes block the capacity of $(-)$-deprenyl to increase the survival of the trophically-withdrawn PC12 cells and the axotomized immature facial motoneurons but do not increase the death of either cell type when delivered without $(-)$-deprenyl (Ansari, Zhang, Holland and Tatton, unpublished findings). The P450 blockers do not alter the capacity of $(-)$-desmethyldeprenyl, the other major metabolite of $(-)$-deprenyl, to increase cell survival and that capacity is similar to that of $(-)$-deprenyl (Tatton and Holland, unpublished observations). These studies of $(-)$-deprenyl metabolites suggest that $(-)$-desmethyldeprenyl, rather than $(-)$-deprenyl, is the active molecule in inducing a decrease in neuronal apoptosis and that in situations in which brain $(-)$-methamphetamine or $(-)$-amphetamine are present at high levels, the capacity of $(-)$-deprenyl to reduce neuronal apoptosis would be markedly compromised (see Oh et al., 1994).

We used translational or transcriptional blockade with cycloheximide, actionomycin or camptothecin to show that the increased survival of trophically-deprived PC12 cells depended on new protein synthesis induced by $(-)$-deprenyl (Tatton et al., 1994). Furthermore, kinetic experiments showed that alterations in gene transcription occurring by 4 hours and changes in mRNA translation occurring by 6 hours after $(-)$-deprenyl addition were necessary to increase the survival of the trophically-deprived PC12 cells. Other recent studies in our laboratory using differential display polymerase chain reaction (PCR) and ^{35}S autoradiographs of two dimensional protein gels have shown that $(-)$-deprenyl alters the transcription/synthesis of 40+ genes/proteins in the trophically-deprived PC12 cells as they enter into apoptosis (Ju and Tatton, unpublished observations). To date, seven of those genes/proteins have been identified using both reverse transcription PCR and western blots: Cu/Zn superoxide dismutase (SOD1), Mn superoxide dismutase (SOD2), tyrosine hydroxlase, Bcl-2, Bax, c-Jun and c-Fos. Most genes/proteins, for example neurofilament protein and tubulin, were not effected by $(-)$-deprenyl showing that the transcriptional changes are selective. Overexpression of Bcl-2 makes a variety of neurons resistant to insults that induce apoptosis while overexpression of Bax increases vulnerability to apoptosis (see Oltvai and Korsmeyer, 1994). Overexpression of SOD1 in nerve cells has been shown to decrease apoptosis, presumably by preventing increased oxidative radical levels which appear to form a critical step in the progression of apoptosis (Greenlund et al., 1995). Maintenance of tyrosine hydroxylase is essential in maintaining catecholamine synthesis. c-Fos and c-Jun are known to contribute to early events in changes in gene expression (Robertson et al., 1995).

Bcl-2 inserts in the outer membranes of mitochondria and is also found in nuclear membranes and the membranes of the endoplasmic reticulum (Lithgow et al., 1994). At an early stage in apoptosis, mitochondria lose their transmembrane potential (MMP) (Vayssiere et al., 1994) and that loss correlates with a loss of mitochondrial ATP production (Richter and Kass, 1991; Bernardes et al., 1994). Overexpression of the cDNA for bcl-2 prevents the loss of MMP in apoptotic cells (Hennet et al., 1993). The capacity of bcl-2 overexpression to block apoptosis is overridden by mitochondrial dysfunction caused by inhibitors of the mitochondrial respiratory chain while those inhibitors can induce apoptosis in cells that express normal levels of Bcl-2 (Smets et al., 1994; Wolvetang et al., 1994). Finally, Bcl-2 overexpression has been shown to decrease the increased oxidative radical levels associated with apoptosis (Hockenbery et al., 1993), possibly by decreasing the oxidative radical formation caused by mitochondrial failure.

We therefore used the potentiometric dyes, chloromethyl tetramethylrosamine (CMTMR), a rhodamine derivative, and 5,5′,6,6′-tetraethylbenzimidazolo carbocyanine (JC-1), a carbocyanine derivative, with fluorescence and confocal microscopy to examine the relationship between PC12 cell apoptosis and mitochondrial failure (Wadia and Tatton, unpublished findings). A significant reduction in MMP was apparent between 3 and 6 hours after trophic withdrawal and increased progressively at 12 and 24 hours. (−)-Deprenyl (10-9M) prevented a reduction in MMP in the PC12 cells at all time points after trophic withdrawal (Wadia and Tatton, unpublished observations).

We then used the fluorochrome RHOD-AM to measure mitochondrial free Ca^{2+} levels. At 6 and 12 hours after trophic withdrawal, mitochondrial Ca^{2+} levels were significantly increased in the PC12 cells while addition of (−)-deprenyl significantly lowered free Ca^{2+} levels in the mitochondria but did not return to the levels found in PC12 cells supported by serum and NGF (Wadia and Tatton, unpublished observations). The maintenance of MMP and therefore mitochondrial energy production by (−)-deprenyl therefore seems to depend at least in part on a capacity to reduce intramitochondrial Ca^{2+} levels. However, since the relationship between intramitochondrial Ca^{2+} levels and MMP is different from that found in PC12 cells supported by serum and NGF, the levels of other intramitochondrial ions which contribute to MMP must also be altered by (−)-deprenyl.

Finally, we used dichlorofluorescein fluorescence to estimate oxidative radical levels in the PC12 cell cytosol after trophic withdrawal. The studies have shown that oxidative radical levels are increased at 12 and 24 hours after trophic withdrawal and that (−)-deprenyl significantly reduces those increases toward the levels in PC12 cells supported by serum and NGF (Wadia and Tatton, unpublished observations). Therefore the progressive reduction in MMP caused by trophic withdrawal is accompanied by an increase in the cytosolic levels of oxidative radicals.

The decrease in cytosolic oxidative radicals seems to indicate that a maintenance of MMP and therefore ATP production induced by the maintenance of Bcl-2 levels found with (−)-deprenyl treatment would result in decreased oxidative radical production by the failing mitochondria. Coupled with increased radical scavenging due to the increases in SOD1 and SOD2 levels induced by (−)-deprenyl, the increased synthesis of the three proteins may be sufficient to account for the decrease in oxidative radicals in the PC12 cells entering apoptosis. Our results may explain the decrease in oxidative radicals with concentrations of (−)-deprenyl too small to inhibit MAO-B found by others in vivo (Wu et al., 1993).

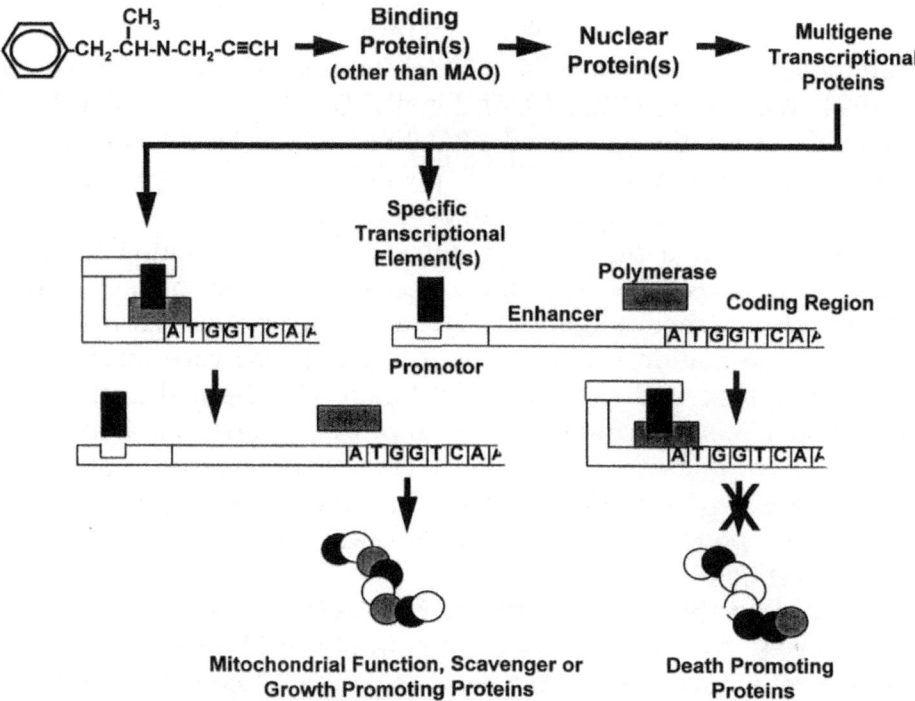

Fig. 1. Schematic for the possible action of (−)-deprenyl in reducing neuronal apoptosis and process growth through selective alterations in transcription. The schematic shows (−)-desmethyldeprenyl acting through a non-MAO binding site to influence specific transcriptional elements to "turn on" the expression of some genes and to "turn off" the expression of others. It hypothesizes that the MAO-independent action of (−)-deprenyl depends in part on mitochondrial proteins, scavenger proteins and growth associated proteins

Conclusions

We propose: 1) that a non-MAO binding site for (−)-desmethyldeprenyl mediates a reduction of neuronal apoptosis and possibly increases in process growth by inducing selective alterations in gene expression (see summary schematic in Fig. 1),

2) that actions of (−)-deprenyl in both neurons and glia require sufficient time for the induction of new protein synthesis and likely only a small proportion of the proteins whose synthesis is altered by (−)-deprenyl have been identified to date,

3) that (−)-deprenyl blocks the mitochondrial failure caused by trophic withdrawal and also increases radical scavenging which together serve to reduce the levels of oxidative radicals known to promote the progress of apoptosis, and

4) that the MAO-B independent transcriptional actions of (−)-deprenyl should be considered as a basis for the apparent slowing of PD and AD observed with the drug.

References

Amenta F, Bongrani S, Cadel S, Ferrante F, Valsecchi B, Zeng YC (1994) Influence of treatment with L-deprenyl on the structure of the cerebellar. Mech Ageing Dev 75: 157–167

Ansari KS, Yu PH, Kruck TP, Tatton WG (1993) Rescue of axotomized immature rat-facial motoneurons by R(−)-deprenyl: stereospecificity and independence from monoamine oxidase inhibition. J Neurosci 13: 4042–4053

Ansari KS, Zhang F, Holland DH, Yu PH, Tatton WG (1993) R(−)-deprenyl, not its major metabolites, rescue axotomized immature facial motoneurons. Soc Neurosci Abstr 19: 243

Bach AWJ, Lan NC, Johnson DL, Abell CW, Bembenek ME, Kwan S-E, Seeburg PH, Shih JC (1988) cDNA cloning of human liver monoamine oxidase A and B: molecular basis of differences in enzymatic properties. Proc Natl Acad Sci USA 85: 4934–4938

Barber AJ, Paterson IA, Gelowitz DL, Voll CL (1993) Deprenyl protects rat hippocampal pyramidal cells from ischemic insult. Soc Neurosci Abstr 19: 1646

Ben-Shachar D, Youdim MB (1991) Intranigral iron injection induces behavioral and biochemical "parkinsonism" in rats. J Neurochem 57: 2133–2135

Berkelaar M, Clarke DB, Wang YC, Bray GM, Aguayo AJ (1994) Axotomy results in delayed death and apoptosis of retinal ganglion cells in adult rats. J Neurosci 14: 4368–4374

Bernardes CF, Meyer-Fernandes JR, Basseres DS, Castilho RF, Vercesi AE (1994) $Ca(2+)$-dependent permeabilization of the inner mitochondrial membrane by 4,4′-diisothicyanatostilbene-2,2′-disulfonic acid (DIDS). Biochim Biophys Acta 1188: 93–100

Berry MD, Scarr E, Zhu MY, Paterson IA, Juorio AV (1994) The effects of administration of monoamine oxidase-B inhibitors on rat striatal neurones response to deprenyl. Br J Pharmacol 113: 1159–1166

Biagini G, Zoli M, Fuxe K, Agnati LF (1993) L-Deprenyl increases GFAP immunoreactivity selectively in activated astrocytes in rat brain. NeuroRep 4: 955–958

Biagini G, Frasoldati A, Fuxe K, Agnati LF (1994) The concept of astrocyte-kinetic drug in the treatment of neurodegenerative diseases: evidence for L-deprenyl-induced activation of reactive astrocytes. Neurochem 25: 17–22

Birkmayer W, Riederer P (1984) Deprenyl prolongs the therapeutic efficacy of combined l dopa in parkinsons disease. Adv Neurol 40: 0–89004

Birkmayer W, Riederer P, Ambrozi L, Youdim MBH (1977) Implications of combined treatment with "madopar" and L-deprenyl in Parkinson's disease. Lancet i: 439–443

Brannan T, Yahr MD (1995) Comparative study of selegiline plus L-dopa-carbidopa versus L-dopa-carbidopa alone in the treatment of Parkinson's disease. Ann Neurol 37: 95–98

Butcher SP, Fairbrother IS, Kelly JS, Arbuthnott GW (1990) Effects of selective monoamine oxidase inhibitors on their in vivo release and metabolism of dopamine in the rat striatum. J Neurochem 55: 981–988

Buys YM, Trope GE, Tatton WG (1995) (−)-Deprenyl increases the survival of retinal ganglion cells after optic nerve crush. Curr Eye Res 14: 119–126

Carlsson A, Fornstedt B (1991) Possible mechanisms underlying the special vulnerability of dopaminergic neurons. Acta Neurol Scand 84: 16–18

Carrillo M, Kanai M, Nokubo M, Kitani K (1990) (−)-Deprenyl induces activities of both superoxide dismutase and catalase but not of glutathione peroxidase in the striatum of young male rats. Life Sci 48: 517–521

Carrillo MC, Kanai S, Nokubo M, Ivy GO, Sato Y, Kitani K (1992) (−)-Deprenyl increases activities of superoxide dismutase and catalase in striatum but not in hippocampus — the sex and age-related differences in the optimal dose in the rat. Exp Neurol 116: 286–294

Carrillo MC, Kitani K, Kanai S, Sato Y, Miyasaka K, Ivy GO (1994) (−)-Deprenyl increases activities of superoxide dismutase and catalase in the rat brain. Life Sci 54: 975–981

Chiueh CC, Huang SJ, Murphy DL (1994) Suppression of hydroxyl radical formation by MAO inhibitors: a novel possible neuroprotective mechanism in dopaminergic neurotoxicity. J Neural Transm [Suppl] 41: 189–196

Cohen G, Spina MB (1989) Deprenyl suppresses the oxidant stress associated with increased dopamine turnover. Ann Neurol 26: 689–690

Dedman DJ, Treffry A, Candy JM, Taylor GAA, Morris CM, Bloxham CA, Perry RH, Edwardson JA, Harrison PM (1992) Iron and aluminium in relation to brain ferritin in normal individuals and alzheimer's disease and chronic renal-dialysis patients. Biochem J 287: 509–514

Demarset KT, Smith DJ, Azzaro AJ (1980) The presence of the type a form of mono amine oxidase ec-1.4.3.4 within nigro striatal dopamine containing neurons. J Pharmacol Exp Ther 215: 461–468

Dluzen DE, McDermott JL (1991) The effect of long term treatment with deprenyl on basal and L-dopa evoked dopamine release in vitro from the corpus striatum of rats. J Neural Transm 85: 145–156

Ekblom J, Jossan SS, Ebendal T, Soderstrom S, Oreland L, Aquilonius S-M (1993) Expression of mRNAs for neurotrophins and members of the Trk family in the rat brain after treatment with L-deprenyl. Neurol Scand 84 [Suppl]: 79–86

Finnegan KT, Skratt JJ, Irwin I, DeLanney LE, Langston JW (1990) Protection against DSP-4 induced neurotoxicity by deprenyl is not related to its inhibition of MAO B. Eur J Pharmacol 184: 119–126

Gerlach M, Riederer P, Youdim MBH (1992) The molecular pharmacology of L-deprenyl. Eur J Pharmacol Mol Pharmacol Sect 226: 97–108

Gerlach M, Ben SD, Riederer P, Youdim MB (1994) Altered brain metabolism of iron as a cause of neurodegenerative diseases? J Neurochem 63: 793–807

Glover V, Sandler M, Owen F (1977) Dopamine is a monoamine oxidase B substrate in man. Nature 265: 80–81

Good PF, Olanow CW, Perl DP (1992) Neuromelanin-containing neurons of the substantia-nigra accumulate iron and aluminum in Parkinson's disease — a LAMMA study. Brain Res 593: 343–346

Greenlund LJS, Deckwerth TL, Johnson EM (1995) Superoxide dismutase delays neuronal apoptosis: a role for reactive oxygen species in programmed neuronal death. Neuron 14: 303–315

Grimsby J, Chen K, Wang L-J, Lan NC, Shih JC (1991) Human monoamine oxidase A and B genes exhibit identical exon-intron organization. Proc Natl Acad Sci USA 88: 3637–3641

Halliwell B (1992) Reactive oxygen species and the central nervous system. J Neurochem 59: 1609–1623

Heikkila RE, Manzino L, Cabbat FS, Duvoisin RC (1984) Protection against the dopaminergic neurotoxicity of 1-methyl-4-phenyl-1,2,5,6-tetrahydropyridine by monoamine oxidase inhibitors. Nature 311: 467–469

Henderson JT, Seniuk NA, Richardson PM, Gaudie J, Roder JC (1994) Systemic administration of ciliary neurotrophic factor induces cachexia in rodents. J Clin Invest 93: 2632–2638

Hennet TG, Bertoni G, Richter C, Peterhans E (1993) Expression of Bcl-2 protein enhances the survival of mouse fibrosarcoid cells in tumor necrosis factor-mediated cytotoxicity. Cancer Res 53: 1456–1460

Hockenbery DM, Ottval ZN, Xiao-Ming Y, Korsmeyer SJ (1993) Bcl-2 functions in an antioxidant pathway to prevent apoptosis. Cell 75: 241–251

Iwasaki Y, Ikeda K, Shoijima T, Kobayashi T, Tagaya N, Kinoshita M (1994) Deprenyl enhances neurite outgrowth in cultured rat spinal ventral horn neurons. J Neurol Sci 125: 11–13

Jellinger K, Kienzl E, Rumpelmair G, Riederer P, Stachelberger H (1992) Iron-melanin complex in substantia nigra of parkinsonian brains: an x-ray microanalysis. J Neurochem 59: 1168–1171

Johnston JP (1968) Some observations upon a new inhibition of monoamine oxidase in brain tissue. Biochem Pharmacol 17: 431–435

Ju WJH, Holland DP, Tatton WG (1994) (−)-Deprenyl alters the time course of death of axotomized facial motoneurons and the hypertrophy of neighboring astrocytes in immature rats. Exp Neurol 126: 233–246

Karoum F, Chuang LW, Eisler T, Calne DB, Liebowitz MR, Quitkin FM, Klein DF, Wyatt RJ (1982) Metabolism of (−)-deprenyl to amphetamine and methamphetamine may be responsible for deprenyl's therapeutic benefit: a biochemical assessment. Neurology 32: 503–509

Knoll J (1988) The striatal dopamine dependency of life span in male rats. Longevity study with (−)-deprenyl. Mech Ageing Dev 46(1–3): 237–62

Knoll J (1992) The pharmacological profile of (−)-deprenyl (selegiline) and its relevance for humans — a personal view. Pharmacol Toxicol 70: 317–321

Knoll J, Mngjar K (1972) Some puzzling pharmacological effects of monoamine oxidase inhibitors. Adv Biochem Psychopharmacol 5: 393–408

Konradi C, Kornhuber L, Froelich L, Fritze J, Heinsen H, Beckmann H, Sculz E, Riederer P (1989) Demonstration of monoamine oxidase-A and -B in the human brainstem by a histochemical technique. Neuroscience 33: 383–400

Koutsiliere E, O'Callaghan JFX, Chen T-S, Riederer P, Rausch W-D (1994) Selegiline enhances survival and neuritie outgrowth of MPP-treated dopaminergic neurons. Eur J Pharmacol 269: R3–R4

Langston JW, Irwin I, Langston EB, Forno LS (1984) Pargyline prevents 1-methyl-4-phenyl-1, 2, 3, 6-tetrahydrophyridine-induced parkinsonism in primates. Science 225: 1480–1482

Li XM, Juorio AV, Paterson IA, Zhu MY, Boulton AA (1992) Specific irreversible monoamine oxidase-B inhibitors stimulate gene expression of aromatic L-amino acid decarboxylase in PC12-cells. J Neurochem 59: 2324–2327

Lithgow T, van Driel R, Bertram JF, Strasser A (1994) The protein product of the oncogene bcl-2 is a component of the nuclear envelope, the endoplasmic reticulum, and the outer mitochondrial membrane. Cell Growth Differ 5: 411–417

Mangoni A, Grassi MP, Frattola L, Piolti R, Bassi S, Motta A, Marcone A, Smirne S (1991) Effects of mao-b inhibitor in the treatment of Alzheimer disease. Eur Neurol 31: 100–107

Martignoni E, Bono G, Blandini F, Sinforiani E, Merlo P, Nappi G (1991) Monoamines and related metabolite levels in the cerebrospinal fluid of patients with dementia of alzheimer type influence of treatment with L deprenyl. J Neural Transm [PD Sect] 3: 15–26

Mayanil CSK, Baquer NK (1984) Comparison of the properties of semipurified mitochondrial and cytosolic monoamine oxidases from rat brain. J Neurochem 43: 906–912

Mochizuki H, Nakamura N, Nishi K, Mizuno Y (1994) Apoptosis is induced by 1-methyl-4-phenylpyridinium ion (MPP+) in ventral mesencephalic-striatal co-culture in rat. Neurosci Lett 170: 191–194

Mytilineou C, Cohen G (1985) Deprenyl protects dopamine neurons from the neurotoxic effect of 1-methyl-4-phenylpyridinium ion. J Neurochem 45: 1951–1953

Oh C, Murray B, Bhattacharya N, Holland D, Tatton WG (1994) (−)-Deprenyl alters the survival of adult facial motoneurons after axotomy: increases in vulnerable C57BL strain but decreases in Mnd mutants. J Neurosci Res 38: 64–74

Olanow CW (1990) Oxidative reactions in Parkinson's disease. Neurology 40: 32–37

Olanow CW, Hanser PA, Gauger L, Malapira T, Koller W, Hubble J, Bushenbark K, Lilienfeld D, Esterlitz J (1995) The effect of deprenyl and levodopa on the progression of the progression of Parkinson's disease. Ann Neurol 38: 771–777

Oltvai ZN, Korsmeyer SJ (1994) Checkpoints of dueling dimers foil death wishes. Cell 79: 189–192

Parkinson Study Group (1993) Effects of tocopherol and deprenyl on the progression of disability in early Parkinson's disease. NEJM 328: 176–183

Paterson IA, Jurorio AV, Boulton AA (1990) Possible mechanism of action of deprenyl in parkinsonism. Lancet 336: 183

Richardson JS, Subbarao KV, Ang LC (1992) On the possible role of iron-induced free radical peroxidation in neural degeneration in Alzheimer's disease. Ann NY Acad Sci 648: 326–327

Richter C, Kass GEN (1991) Oxidative stress in mitochondria: its relationship to cellular Ca2+ homeostasis, cell death, proliferation and differentiation. Chem Biol Interact 77: 1–23

Robertson LM, Kerppola TK, Smeyne RJ, Bocchairo C, Morgan JI, Curran T (1995) Regulation of c-fos expression in transgenic mice requires multiple interdependent transcription control elements. Neuron 14: 241–252

Rosenbaum DM, Michaelson M, Batter DK, Doshi P, Kessler JA (1994) Evidence for hypoxia-induced, programmed cell death of cultured neurons. Ann Neurol 36: 864–870

Roy E, Bedard PJ (1993) Deprenyl increases survival of rat foetal nigral neurones in culture. Neuroreport 4: 1183–1186

Rukenstein A, Rydel RE, Greene LA (1991) Multiple agents rescue PC12 cells from serum-free cell death by translation- and transcription-independent mechanisms. J Neurosci 11: 2552–2563

Salo PT, Tatton WG (1992) Deprenyl reduces the death of motoneurons caused by axotomy. J Neurosci Res 31: 394–400

Schulzer M, Mak E, Calne DB (1992) The anitparkinsonism efficacy of deprenyl derives from transient improvement which is likely to be symptomatic. Ann Neurol 32: 795–798

Sengstock GJ, Olanow CW, Dunn AJ, Arendash GW (1992) Iron induces degeneration of nigrostriatal neurons. Brain Res Bull 28: 645–649

Seniuk NA, Henderson JT, Tatton WG, Roder JC (1994) Increased CNTF gene expression in process bearing astrocytes following injury is augmented by R(−)-deprenyl. J Neurosci Res 37: 278–286

Shih JC (1991) Molecular basis of human MAO A and B. Neuropsychopharm 4: 1–7

Smets LA, Van denBerg J, Acton D, Top B, Van Rooij H, Verwijs-Janssen M (1994) BCL-2 expression and mitochondrial activity in leukemic cells with different sensitivity to glucocorticoid-induced apoptosis. Blood 84: 1613–1619

Sofic E, Paulus W, Jellinger K, Riederer P, Youdim MB, Amit T (1991) Selective increase of iron in substantia nigra zona compacta of parkinsonian brains. J Neurochem 56: 978–982

Spina MB, Cohen G (1989) Dopamine turnover and glutathione oxidation: implications for Parkinson disease. Proc Natl Acad Sci USA 86: 1398–1400

Tariot PN, Sunderland T, Weingartner H, Murphy DL, Welkowtiz JA, Thompson K, Cohen RM (1987) Cognitive effects of l deprenyl in alzheimer's disease. Psychopharmacology 91: 489–495

Tatton WG, Greenwood CE (1991) Rescue of dying neurons: a new action for deprenyl in MPTP parkinsonism. J Neurosci Res 30: 666–627

Tatton WG, Salo PT, Yu PH, Holland DP, Kwan MM, Ansari KS (1992) Both selegiline and pargyline reduce the death of immature motoneurons caused by axotomy. Proc Soc Neurosci 18: 47

Tatton WG, Seniuk NA, Ju WYH, Ansari KS (1993) Reduction of nerve cell death by deprenyl without monoamine oxidase inhibition. Monoamine oxidase inhibitors in neurological diseases. Raven Press, New York, pp 217–248

Tatton WG, Ju WYL, Holland DP, Tai CE, Kwan MM (1994) (−)-Deprenyl reduces PC12 cell apoptosis by inducing new protein synthesis. J Neurochem 63: 1572–1574

Taylor KM, Snyder SH (1974) Amphetamine: differentiation by D and L isomers of behavior involving brain norepinephrine or dopamine. Science 168: 519–521

Tetrud JW, Langston JW (1989) The effect of deprenyl (selegiline) on the natural history of Parkinson's disease. Science 245: 519–522

Thiffault C, Aumont N, Quirion R, Poirier J (1994) Antioxidant enzymes in an animal model of Parkinson's disease. Can J Physiol Pharmacol 72: 592

Vayssiere JL, Petit PX, Risler Y, Mignotte B (1994) Commitment to apoptosis is associated with changes in mitochondrial biogenesis and activity in cell lines conditonally immortalized with simian virus 40. Proc Natl Acad Sci USA 91: 11752–11760

Vincent SR (1989) Histochemical localization of 1-methyl-4-phenyl-1,2,3,6-tetrahydropyridine oxidation in the mouse brain. Neuroscience 28: 189–199

Waldmeier PC, Felner AE (1978) Deprenil, loss of selectivity for inhibition of b-type MAO after repeated treatment. Biochem Pharmacol 27: 801–806

Westlund KN, Denney RM, Kochersperger LM, Rose RM, Abell CW (1985) Distinct monoamine oxidase a and b populations in primate brain. Science (Washington DC) 230: 181–183

Westlund KN, Denney RM, Rose RM, Abell CW (1988) Localization of distinct monoamine oxidase a and monoamine oxidase b cell populations in human brainstem. Neuroscience 25: 439–456

Wilcox BJ, Clatterbuck RE, Price DL, Koliatsus VE (1993) Evidence for programmed cell death in motor neurons in the rat central nervous sustem (CNS). Soc Neurosci Abstr 19: 441

Wolvetang EF, Johnson KL, Krauer K, Ralph SJ, Linnane AW (1994) Mitochondrial respiratory chain inhibitors induce apoptosis. FEBS Lett 339: 40–44

Wu R-M, Chiueh CC, Pert A, Murphy DL (1993) Apparent antioxidant effect of L-deprenyl on hydroxyl radical formation and nigral injury elicited by MPP+ in vivo. Eur J Pharmacol 243: 241–248

Wu RM, Murphy DL, Chiueh CC (1995) Neuronal protective and rescue effects of deprenyl against MPP+ dopaminergic toxicity. J Neural Transm 100: 53–61

Youdim MB, Ben-Shachar D, Riederer P (1994) The enigma of neuromelanin in Parkinson's disease substantia nigra. J Neural Transm [Suppl] 43: 113–22

Youdim MB, Lavie L, Riederer P (1994) Oxygen free radicals and neurodegeneration in Parkinson's disease: a role. Ann NY Acad Sci 738: 64–68

Yu PH, Davis BA, Fang J, Boulton AA (1994) Neuroprotective effects of some monoamine oxidase-B inhibitors against. J Neurochem 63: 1820–1828

Zhang F, Richardson PM, Holland DP, Guo G, Tatton WG (1995) CNTF or (-)-deprenyl in immature rats: survival of axotomized facial motoneurons and weight loss. J Neurosci Res 40: 564–570

Zhang X, Zuo D-M, Yu PH (1995) Neuroprotection by R(-)-deprenyl and N-2-hexyl-N-methylpropargylamine on DSP-4, a neurotoxin, induced degeneration of noradrenergic neurons in the rat locus coeruleus. Neurosci Lett 186: 45–48

Authors' address: Dr. W. G. Tatton, Institute for Neuroscience, 12th Floor, Tupper Medical Building, Dalhousie University, Halifax, Nova Scotia, Canada

J Neural Transm (1996) [Suppl] 48: 61–73
© Springer-Verlag 1996

Are metabolites of l-deprenyl (selegiline) useful or harmful? Indications from preclinical research

S. Yasar[1,2], **J. P. Goldberg**[3], and **S. R. Goldberg**[2]

[1] Department of Anesthesiology and Critical Care Medicine, Johns Hopkins University, Medical School, Baltimore, MD, [2] Behavioral Pharmacology and Genetics Section, Preclinical Pharmacology Laboratory, Division of Intramural Research, National Institute on Drug Abuse, National Institutes of Health, Baltimore, MD, and [3] Vista Hill Hospital, Chula Vista, CA, USA

Summary. A frequent topic of controversy has been whether metabolism of l-deprenyl (selegiline) to active metabolites is a detriment to clinical use. This paper reviews possible roles of the metabolites of l-deprenyl in producing unwanted adverse side effects or in augmenting or mediating its clinically useful actions. Levels of l-amphetamine and l-methamphetamine likely to be reached, even with excessive intake of l-deprenyl, would be unlikely to produce neurotoxicity and there is no preclinical or clinical evidence of abuse liability of l-deprenyl. In contrast, there is evidence that l-amphetamine and l-methamphetamine have some qualitatively different actions than their d-isomer counterparts on EEG and cognitive functioning which might result in beneficial clinical effects and complement beneficial clinical actions of l-deprenyl itself.

Introduction

l-Deprenyl is a useful and effective drug in the clinical treatment of parkinsonism and holds promise for treatment of Alzheimer's disease and other neurodegenerative diseases. However, since it was originally developed as a "psychic energizer" antidepressant (Knoll et al., 1965), has an amphetamine-like phenylethylamine structure and undergoes metabolic transformation to active compounds, with it's major metabolites in vivo being l-methamphetamine and l-amphetamine (Reynolds et al., 1978; Phillips, 1981; Elsworth et al., 1982; Karoum et al., 1982), there has been a recurrent concern with its use (e.g., Goldberg et al., 1994). Since l-amphetamine and presumably l-methamphetamine, like their d-stereoisomers, release dopamine and norepinephrine from presynaptic terminals (Heikkila et al., 1975; Engberg et al., 1991; Fang and Yu, 1994; Kuczenski et al., 1995) and have psychomotor stimulant-like behavioral effects (e.g., Yokel and Pickens, 1973; Katz, 1982; Kuczenski et al., 1995), these metabolites may play a role in l-deprenyl's neuropharmacological and behavioral effects (see discussions by Fozard et al.,

1985; Engberg et al., 1991; Berry and Paterson, 1994), particularly when the dose of l-deprenyl is increased above clinically-relevant levels.

The clinical use of amphetamine-like psychomotor stimulants is limited by their potential for abuse, and the l-stereoisomers of amphetamine and meth-amphetamine are schedule II controlled substances (Part 1308 of the U.S. Controlled Substances Act) with apparent abuse liability (e.g., Cody and Schwarzhoff, 1993). They are available only for prescription use, with the exception of l-methamphetamine's use in Vick's inhaler which is exempted from control. The question naturally arises as to whether l-deprenyl's metabo-lites would confer upon it amphetamine-like abuse liability. Also, the l-stereoisomers of amphetamine and methamphetamine have a variety of other neuropharmacological actions including effects on EEG and behavior that might contribute to clinical actions of l-deprenyl. Thus, it is of interest to review the possible roles of the metabolites in producing unwanted adverse side effects of l-deprenyl or in augmenting or mediating clinically useful actions of l-deprenyl.

Behavioral actions of l-deprenyl and its metabolites

Over the past 30 years objective and quantitative techniques for establishment of reproducible behavioral baselines in laboratory animals have been devel-oped which are sensitive to the effects of psychoactive drugs (e.g., Goldberg and Stolerman, 1986). For example, laboratory animals can be trained to differentially respond between a drug and its vehicle by requiring them to emit one response following the administration of that drug and another response following the administration of the drug vehicle, with the discrimina-tion being made presumably on the basis of drug-induced interoceptive stimuli. Stimuli such as administration of psychoactive drugs which set the occasion for differential responding are termed "discriminative stimuli" (e.g., Terrace, 1966). Also, abuse liability of drugs can be directly assessed in laboratory animals through the use of i.v. drug self-administration procedures. The most frequently used procedure involves training monkeys or rats to self-administer i.v. injections of a prototype drug from a class of drugs with abuse potential (e.g. cocaine for psychomotor stimulants) and then testing the drug in question by substitution.

Drug-discrimination studies

Drug discrimination techniques in experimental animals have provided a powerful tool for examining the interoceptive-stimulus effects of psychomo-tor stimulants such as amphetamine and cocaine and for investigating neuro-chemical mechanisms that underlie their interoceptive-stimulus actions. The use of drug-discrimination techniques has been successfully extended to ex-perimental studies of human subjects and these techniques are thought to provide an objective means of measuring subjective reports of quality and strength of drug effect. A number of investigators have used the techniques to evaluate the ability of l-deprenyl to produce amphetamine- or cocaine-like interoceptive-stimulus effects in animals (see review by Yasar et al., 1993b). Interestingly, l-deprenyl had been found to fully substitute as a discriminative stimulus for both d-amphetamine and l-cocaine in rats (Colpaert et al., 1980; Porsolt et al., 1984; Moser, 1990; Yasar et al., 1993a,b, 1994; Yasar and Bergman, 1994) and for d-methamphetamine in monkeys (Yasar and Bergman, 1994) and it appears that these discriminative stimulus effects of l-deprenyl are mediated by its metabolites.

Drug-discrimination studies commonly employ two-lever, operant, choice procedures. During daily experimental sessions, a rat or monkey is placed in an operant chamber (rat) or chair (monkey), which contains two levers, stimulus lights and, when food is used as a reinforcer, an opening to a receptacle for small food pellets. In addition, electric shock can be delivered either through metal rods comprising the floor of the chamber (rats) or through electrodes placed on the tail (monkey). For lever-pressing responses to be reinforced during training, the animal must press the lever appropriate (as defined by the investigator) to the presession drug condition. For example, in order for lever presses to produce food pellets, the rat might be required to press the right lever on days when they receive an amphetamine injection but press the left lever on days when they receive vehicle or no drug. Thus, training is accomplished by reinforcing only correct responses and by not reinforcing incorrect responses. Usually drug and no-drug days are alternated until the subject reaches a defined criterion of making a certain percentage of responses during the session on the correct lever (e.g., 90%). When this criterion performance is reached, test sessions are then alternated with training sessions. During test sessions, responding on either lever is usually considered correct. Different doses of the training drug are usually tested first and a range of doses of other compounds are then tested for generalization to the training drug. Dose-response functions generated in this way relate responding to drug dose and are termed "stimulus generalization curves".

In a series of experiments by Yasar and colleagues (Yasar et al., 1993a,b; Yasar and Bergman, 1994) male Fischer rats were trained under a 5-response fixed-ratio (FR5) schedule of stimulus-shock termination to respond on one lever following i.p. injections of 1.0 mg/kg d-amphetamine and on a second lever following i.p. injections of saline-vehicle. Under this schedule, each completion of 5 responses terminated shock-associated visual stimuli and initiated a 45-sec timeout period. Experiments were conducted to determine the effects of a range of doses of d-amphetamine (0.1–3.0 mg/kg), l-amphetamine (0.1–2.0 mg/kg), l-deprenyl (2.0–30.0 mg/kg) and d-deprenyl (2.0–17.0 mg/kg). Individual doses of a drug were administered i.p. 30 min (deprenyl) or 15 min (other drugs) prior to the session which lasted 30 min or less. Both enantiomers of amphetamine and deprenyl produced dose-related increases in responding on the d-amphetamine-associated lever. On average, full substitution (≥90% responding on the d-amphetamine-associated lever) was produced by 1.0 mg/kg d-amphetamine, 1.0 mg/kg l-amphetamine, 10.0 mg/kg d-deprenyl and 17.0 mg/kg l-deprenyl (Fig. 1). Rates of responding were not appreciably altered by doses of drugs that substituted for d-amphetamine, and were similar to those observed following injections of

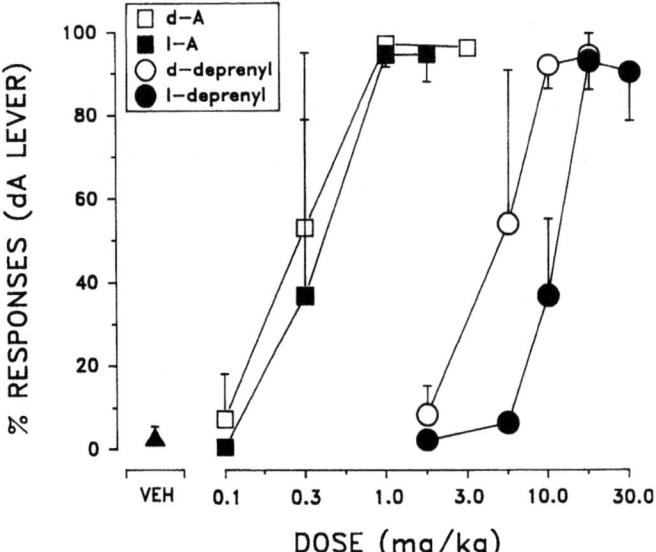

Fig. 1. Effects of d-amphetamine, l-amphetamine d-deprenyl, and l-deprenyl in rats trained to discriminate 1.0 mg/kg of d-amphetamine from saline. *Abscissa*: dose, log scale; *ordinate*: percentage of responses on the d-amphetamine-associated lever. Each point represents the mean (±SEM) of effects in six rats. Data for VEH represent the effects of saline (modified from Yasar et al., 1993a)

saline-vehicle. Similar results were found both when behavior was maintained by food delivery rather than stimulus-shock termination and when rats were trained to discriminate l-cocaine rather than d-amphetamine from saline (Yasar et al., 1994).

These stimulant-like discriminative stimulus actions of l-deprenyl were not simply a result of studying a species (the rat) with markedly different distributions from humans of MAO-B relative to MAO-A activity within the central nervous system (Collins et al., 1970), since similar results were obtained when non-human primates were studied. In another series of experiments by Yasar and Bergman (1994), squirrel monkeys (*Saimiri sciureus*) were trained under a schedule of stimulus, electric-shock termination to respond on one lever following i.m. injection of 0.3 mg/kg d-methamphetamine and on a second lever following i.m. injection of saline-vehicle. Each completion of ten responses (FR10) terminated shock-associated visual stimuli and initiated a 50-sec timeout period. During test sessions, completion of the FR10 schedule on either lever turned off shock-associated visual stimuli and initiated the 50-sec timeout period. Using cumulative-dosing procedures during test sessions, incremental doses of d- or l-methamphetamine were injected i.m. during timeout periods that preceded sequential components. In most experiments, four or five doses of each compound were studied by administering overlapping ranges of cumulative doses during different test sessions. In the d-methamphetamine-trained monkeys, both d- and l-methamphetamine produced dose-related increases in responding on the drug-associated lever

(Fig. 2), with full substitution in all monkeys after cumulative doses of 0.3 mg/ kg d-methamphetamine and 3.0 mg/kg l-methamphetamine.

Subsequent studies with l-deprenyl indicated that it's behavioral effects became evident within the second hour following its i.m. administration (Yasar and Bergman, 1994). Therefore, single doses of l-deprenyl were injected i.m. 60 min prior to the session and only data from the last two components of the session, i.e. 100–130 min following injection, were used for analysis of results. l-Deprenyl was found to substitute for d-methamphetamine in a dose-related manner, with full substitution in all monkeys after administration of a dose of 5.6 mg/kg (Fig. 2). In contrast, other studies showed that the effective dose range for MAO-B inhibition was 0.1 to 0.3 mg/kg (Bergman, unpublished observations). Thus, the dose of l-deprenyl needed to produce full generalization to the methamphetamine discriminative stimulus was 20 to 60 times greater than the clinically relevant doses that effectively inhibited MAO-B activity. Even after this high dose of l-deprenyl, responding in early components of the session generally occurred on the saline-associated lever. By the third component of the session, however, responding occurred predominantly on the drug-associated lever.

The above findings that at very high doses l-deprenyl can substitute for both d-amphetamine and cocaine in rats and for d-methamphetamine in monkeys are consistent with previous reports (Colpaert et al., 1980; Porsolt et al., 1984). Since doses of l-deprenyl that selectively inhibit MAO-B activity and, thus, markedly increase circulating levels of β-PEA in rats (0.1 to 0.5 mg/

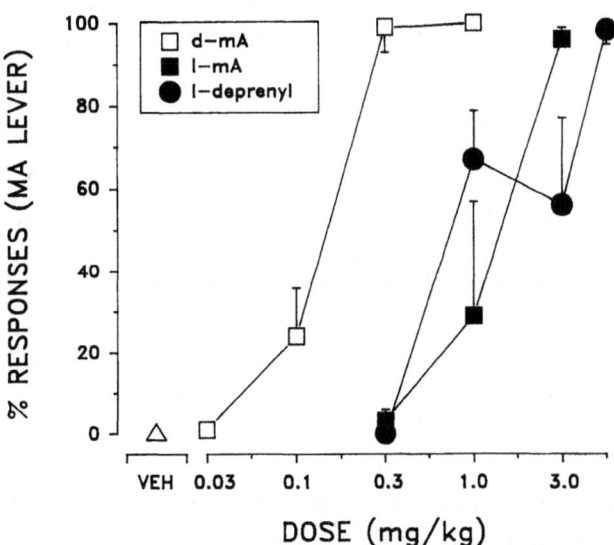

Fig. 2. Effects of d-methamphetamine, l-methamphetamine, and l-deprenyl in monkeys trained to discriminate 0.3 mg/kg d-methamphetamine from saline. *Abscissa*: dose, log scale; *ordinate*: percentage of responses on the d-methamphetamine-associated lever. Each point represents the mean of effects in four squirrel monkeys. Effects of the enantiomers of methamphetamine were determined using cumulative i.m. dosing procedures; the effects of l-deprenyl were determined by administration of single i.m. doses (from Yasar and Bergman, 1994)

kg, i.p. or s.c., Knoll, 1978; Buu and Angers, 1987; Patterson et al., 1991) and monkeys (0.1 to 0.3 mg/kg, i.m., Bergman, unpublished results) are at least 20-fold lower than those that fully substituted for d-amphetamine, l-cocaine or d-methamphetamine in the above experiments by Yasar and colleagues, it would appear that MAO-B inhibition did not play a major role in these effects and that the amphetamine- and cocaine-like discriminative stimulus effects of l-deprenyl are, most likely, due to it's conversion to active metabolites (See Yasar and Bergman, 1994, for further discussion).

Intravenous drug self-administration behavior

Although the doses of l-deprenyl required to produce stimulant-like discriminative stimulus effects in animals are 20- to 60-fold higher than those that would be relevant clinically, increased availability and decreased cost of l-deprenyl that will likely occur when it becomes a generic compound in the near future could renew concern about abuse liability resulting from it's metabolism to l-methamphetamine and l-amphetamine. In order to preclinically assess abuse liability, the reinforcing effects of l-deprenyl have been directly assessed by using i.v. self-administration procedures in monkeys (Yasar et al., 1993b; Winger et al., 1994). A procedure for rapidly evaluating reinforcing effects of intravenously delivered drugs in rhesus monkeys (*macaca mulatta*) was used (Winger et al., 1989). Monkeys with chronic venous catheters were individually housed in cages in which drug was available during two sessions a day, with each session divided into four components; each component ended after 20 injections or after 25 min had passed, and was separated from the following component by a 10-min timeout interval. Thirty lever presses were required to produce each injection and a unique dose per injection of drug was available during each of the four components. Monkeys were initially trained with l-cocaine. Saline was periodically substituted for cocaine until rates of responding during saline-substitution components of the session were consistently below 0.5 responses per second. Different doses of l-cocaine, d-deprenyl, l-deprenyl or l-methamphetamine, the major metabolite of l-deprenyl, were then substituted during single sessions.

Figure 3 shows that under the above condition l-cocaine was a very effective reinforcer, maintaining rates of lever pressing of over 2 responses per second at an injection dose of 0.01 mg/kg. In contrast, neither the d- or the l-isomers of deprenyl were self-administered above saline placebo levels. l-Deprenyl failed to maintain self-administration responding although it was studied over a thousand-fold dosage range from 0.001 to 1.0 mg/kg/injection. It should be noted that at the 1.0 mg/kg injection dose a monkey could receive as much as 20.0 mg/kg i.v. l-deprenyl in about 25 min if responding occurred rapidly, as it did when other drugs such as cocaine with known abuse liability were available.

l-Methamphetamine, the major metabolite of l-deprenyl, was as effective a reinforcer as l-cocaine, maintaining rates of lever pressing of over 2 re-

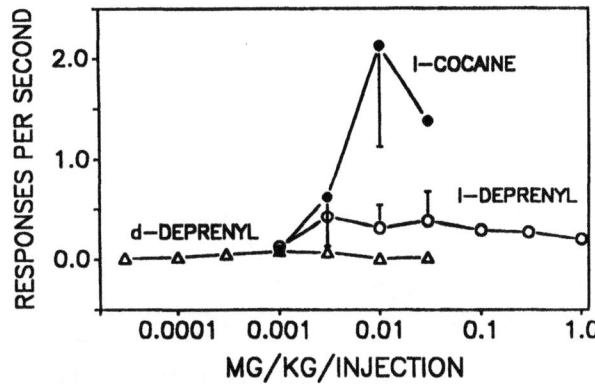

Fig. 3. Rates of lever-press responding maintained by intravenously delivered l-cocaine (solid circles), l-deprenyl (open circles) or d-deprenyl (open triangles) in rhesus monkeys. *Abscissa*: dose, log scale, in mg/kg/injection; *ordinate*: mean rates of responding in responses per second. Where no vertical SEM bars are shown, they are either within the size of the symbol or fewer than three monkeys were studied at that dose (from Winger et al., 1994)

sponses per second at an injection dose of 0.03 mg/kg (Fig. 4). Winger et al. (1994) also tested the effects of a 1.0 mg/kg dose of l-deprenyl on both l-methamphetamine and l-cocaine self-administration to preclinically evaluate l-deprenyl's potential as a medication for treatment of drug abuse. Although the 1.0 mg/kg dose of l-deprenyl tested was as high as they could go without signs of nonspecific toxicity in the monkeys, such as reduction of food intake, there was no effect on either l-methamphetamine (Fig. 4) or l-cocaine (data not shown) self-administration.

A major endogenous substrate for MAO-B, the enzyme specifically and irreversibly inactivated by l-deprenyl, is β-phenylethylamine, an endogenous amine with amphetamine-like actions which normally is rapidly inactivated when injected. At high enough doses, β-phenylethylamine has transient stimulant-like properties and is self-administered (Risner and Jones, 1977; Shannon and De Georgio, 1982). In experiments reviewed by Yasar et al. (1993), a 1.0 mg/kg dose of l-deprenyl that was not self-administered and had no effect on l-methamphetamine or l-cocaine self-administration in the experiments described above, produced a marked shift to the left of the dose-response curve for β-phenylethylamine self-administration. Similar interactions have been reported between l-deprenyl and β-phenylethylamine with other behavioral measures. For example, Timár and Knoll (1986) found that s.c. treatment with 0.25 to 2.0 mg/kg doses of l-deprenyl greatly potentiated the intensity and duration of stereotyped behavior induced by a 40.0 mg/kg s.c. dose of β-phenylethylamine.

To summarize then, l-deprenyl did not serve as a reinforcer of self-administration behavior, even at injection doses that would have allowed the monkeys to self-administer as much as 20 mg/kg of l-deprenyl within a 25-minute session. In contrast, a 1 mg/kg dose of l-deprenyl was clearly in the clinically relevant dose range since it produced a marked shift to the left of the

68   S. Yasar et al.

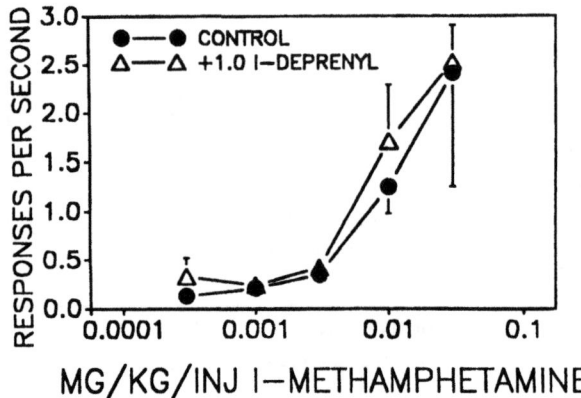

MG/KG/INJ l−METHAMPHETAMINE

Fig. 4. Efects of 1.0 mg/kg l-deprenyl (open triangles) on rates of responding maintained by intravenous self-administration of l-methamphetamine. Rate-maintaining effects of l-methamphetamine in the absence of l-deprenyl are indicated by closed circles. *Abscisa*: dose of l-methamphetamine in mg/kg/injection; *ordinate*: mean rates of responding in responses per second. Where no vertical SEM bars are shown, they are within the size of the symbol (from Winger et al., 1994)

dose-response curve for β-phenylethylamine. It is interesting to note that despite the formation of active metabolites that would occur after pretreatment with l-deprenyl, l-deprenyl pretreatment had no specific effect on either l-methamphetamine or on l-cocaine self-administration.

EEG and cognitive functioning

There is evidence suggesting that the l-amphetamine and l-methamphetamine metabolites of l-deprenyl have nootropic properties (i.e., learning and memory enhancing properties, particularly under impaired brain functioning; Bartus, 1990) that could be additive to those of l-deprenyl. Nickel et al. (1990, 1994) have shown in rats that both l-deprenyl and l-amphetamine increased EEG theta rhythms (3.7–7.5 Hz), in contrast to d-deprenyl and d-amphetamine which decreased theta rhythms. This was not due to a general slowing effect on EEG rhythm since the lower frequency delta rhythm (1.5–3.5 Hz) was, in contrast, decreased by l-deprenyl and l-amphetamine and increased by d-deprenyl and d-amphetamine. Furthermore, increases in theta rhythms occurred at a dose of 1 mg/kg of l-deprenyl, which would be equivalent to a therapeutic dose in humans, as well as at a much higher dose of 5 mg/kg. In contrast, d-amphetamine- or cocaine-like discriminative stimullus effects of l-deprenyl in rats occur at doses of 17 mg/kg, i.p., or more (Yasar et al., 1993a,b, 1994).

Given that theta rhythm generation is an indication in humans that a drug has facilitory actions in areas of the brain related to learning and memory, selective attention, problem solving and memory consolidation (e.g., limbic areas) (Lynch et al., 1990; Miller, 1991; Ramos et al., 1993; Corsi-Cabrera et al., 1993), theta generation by l-amphetamine (and assumedly by l-

Table 1. Changes in the electroencephalographic frequency bands in freely moving rats after oral administration of l-deprenyl, l-amphetamine, d-amphetamine, and d,l-amphetamine in percentages compared with the control value for the same animal

Compounds	Dose mg/kg (by mouth)	Minutes after treatment	δ 1.5–3.5 Hz (%)	θ 3.7–7.5 Hz (%)	α 7.8–13.5 Hz (%)	β 13.8–30.0 Hz (%)
l-Deprenyl	1	60	−12*	+16**	−2	−4
l-Deprenyl	5	60	−22**	+21**	−4	+2
d-Deprenyl	1	60	+14**	−18**	+3	+4
d-Deprenyl	5	60	+21**	−28**	−2	+11*
l-Amphetamine	10	60	−15**	+13**	+5	−8*
d-Amphetamine	1	60	+19**	−20**	−2	+6*
d,l-Amphetamine	5	60	+17**	−21**	+1	+7*

Reprinted from Nickel et al. (1990). *,** Asterisks indicate that the electroencephalographic frequency changes after administration of drug differed significantly from the baseline value obtained before administration of drug ($*p < 0.05$; $**p < 0.01$)

methamphetamine) indicates that these metabolites themselves may facilitate or be responsible for any cognitive properties of l-deprenyl. Since such actions do not appear to be possessed by the d-isomers, any benefits from the l-methamphetamine and l-amphetamine metabolites would not be due to just general activating (i.e., stimulating) d-amphetamine-like actions which have not been shown to have true learning benefits. The increases in theta rhythms after treatment with the l-isomers of deprenyl and amphetamine, in contrast to the decrease in these rhythms after treatment with the d-isomers, indicates a higher and more benign area of stimulation.

There is further evidence that cognitive facilitation by l-deprenyl may be independent of MAO-B inhibition and likely results from formation of active metabolites which are l-isomers of amphetamines (Milgram et al., 1993; Gelowitz et al., 1994). For example, Gelowitz et al. (1994) showed that chronic treatment with low and infrequent doses of either l-amphetamine or l-deprenyl prevented aging-related learning decrements in rats to an equal extent, despite a normal increase in MAO-B activity with aging in l-amphetamine treated rats. Rats were injected s.c. with l-deprenyl (0.25 mg/kg), l-amphetamine (0.25 mg/kg) or saline 3 times a week for 4 months starting when they were 6 months old. At 10 months of age (middle-aged rats) behavioral testing was started while injections continued. Middle-aged l-deprenyl-and l-amphetamine-treated rats and young control rats all learned the water-maze task equally well and all were better than the middle-aged rats given only saline. MAO-B activity was significantly higher in middle-aged rats given either saline or l-amphetamine than in young controls. In contrast, both MAO-B and MAO-A activity were lower in l-deprenyl treated middle-aged rats than in rats in the three other groups. Thus, behavior in older rats was equally well facilitated by either l-deprenyl or l-amphetamine, despite large differences in MAO-B (and MAO-A) activity, indicating that the active l-amphetamine and l-methamphetamine metabolites of l-deprenyl were responsible for learning facilitation.

Discussion

Although conventional stimulants such as d-amphetamine have shown benefits for the treatment of depression in geriatric and/or ill patients, it is not feasible to continue treatment with these stimulants beyond a 2–4 week acute phase due to the possibility of abuse or excessive stimulation. l-Isomers of amphetamine or methamphetamine may have cognitive and mood-enhancing actions without the excessive stimulation of behavior characteristic of their d-isomers. Also, although normal psychostimulants (such as caffeine or d-amphetamine) can increase alertness, vigilance and attention and may speed information processing in elderly adults, they are less likely to correct cognitive deficiencies which occur during aging such as directed problem solving and processing of novel material or information (old dog difficulties in learning new tricks) or accuracy of response (rate of false positives, etc.) (e.g., Halliday et al., 1986; Koelega, 1993). The l-isomers of amphetamine and methamphetamine may have a greater benefit to correct these processing defects. Furthermore, the slow and secondary release of l-deprenyl-produced l-methamphetamine and l-amphetamine metabolites can be expected to minimize or prevent any reinforcing actions of these metabolites (Winger et al., 1994). Additionally, l-amphetamine and l-methamphetamine could have beneficial mood effects not possessed by l-deprenyl itself through their more potent effects on dopamine and norepinephrine uptake and release (e.g., Fang and Yu, 1994; Kuczenski et al., 1995). These properties could have antidepressant, activating actions which might counteract negative mood effects often intrinsic to aging-related illnesses. These uptake actions also might have neuroprotective benefits which complement those possessed by l-deprenyl itself (Marck et al., 1990; Sprague and Nichols, 1995; Tatton, 1993).

It should be noted that unscheduled and commonly used antidepressants such as nomifensine and bupropion, which have marked dopamine-uptake inhibitory actions and also show robust reinforcing actions in both rats and monkeys at doses no greater than twice their equivalent therapeutic dose range (Spyraki and Fibiger, 1981; Lamb and Griffiths, 1990), have a long history of clinical use but have not been shown to cause abuse or dependence in humans. Both these antidepressants contain the phenylethylamine skeleton found in amphetamine and other psychomotor stimulants and, like l-deprenyl, they reliably generalize to cocaine and amphetamines in drug-discrimination studies with rats and monkeys (e.g., Lamb and Griffiths, 1990). In contrast, l-deprenyl does not appear to function as a reinforcer in monkeys and generalizes to cocaine or amphetamines in drug-discrimination studies with rats and monkeys only at doses 20 fold higher than the clinically relevant therapeutic dose range. Winger et al. (1994) have further shown that pretreatment with l-deprenyl does not potentiate or sensitize to the reinforcing actions of l-cocaine despite elevation of β-phenylethylamine levels and generation of l-methamphetamine and l-amphetamine metabolites, both of which should heighten the dopaminergic response to cocaine which has been assumed to underly cocaine's reinforcing actions.

Finally, there may be as yet undemonstrated beneficial clinical actions of l-deprenyl based on its slow release of l-amphetamine and l-methamphetamine, unrelated to treatment of aging-related diseases. For example, cognitive degeneration subsequent to illness or injury (post-stroke, vascular dementia, closed head injury, etc.) typically involves depression in as many as 50% of patients (e.g., Wragg and Jeste, 1989). Psychomotor stimulants are often used to treat depression in such geriatric patients (e.g., Woods et al., 1986; Chiarello and Cole, 1987; Warneke, 1990; Masand et al., 1991). l-Deprenyl, with it's l-amphetamine and l-methamphetamine metabolites might be of use in treating both the cognitive deficiency and depression in such patients and, thus, might serve as maintenance therapy.

To conclude then, it is not clear that the metabolism of l-deprenyl to active metabolites is a detriment to its clinical use. The levels of l-amphetamine and l-methamphetamine likely to be reached, even with excessive intake of l-deprenyl, would be magnitudes lower than those which would produce clear signs of neurotoxicity with the corresponding d-stereoisomers and there is no preclinical or clinical evidence of abuse liability of l-deprenyl resulting from formation of active metabolites. In contrast, there is evidence that the l-stereoisomers of l-amphetamine and l-methamphetamine may have some qualitatively different actions than their d-isomer counterparts which might result in beneficial clinical effects and could complement any beneficial clinical actions of l-deprenyl itself.

References

Bartus RT (1990) Drugs to treat age-related neurodegenerative problems. J Aging Geriatr Sci 38: 680–695

Berry MD, Juorio AV, Paterson IA (1994) Possible mechanisms of action of (−)-deprenyl and other MAO-B inhibitors in some neurologic and psychiatric disorders. Prog Neurobiol 44: 141–161

Buu NT, Angers M (1987) Effects of different monoamine oxidase inhibitors on the metabolism of L-DOPA in the rat brain. Biochem Pharmacol 36: 1731–1735

Chiarello RJ, Cole JO (1987) The use of psychostimulants in general psychiatry. Arch Gen Psychiatry 44: 286–295

Cody JT, Schwarzhoff R (1993) Interpretation of methamphetamine and amphetamine enantiomer data. J Anal Toxicol 17: 321–326

Colpaert FC, Niemegeers CJE, Janssen PAJ (1980) Evidence that a preferred substrate for type B monoamine oxidase mediates stimulus properties of MAO inhibitors: a possible role for β-phenylethylamine in the cocaine clue. Pharmacol Biochem Behav 13: 513–517

Corsi-Cabrera R, Ramos J, Guevara MA, Arce C, Gutierrez S (1993) Gender differences in the EEG during cognitive activity. Int J Neurosci 72: 257–264

Engberg G, Elebring T, Nissbrandt H (1991) Deprenyl (selegiline), a selective MAO-B inhibitor with active metabolites; effects on locomotor activity, dopaminergic neurotransmission and firing rate of nigral dopamine neurons. J Pharmacol Exp Ther 259: 841–847

Fang J, YU PH (1994) Effect of L-deprenyl, its structural analogues and some monoamine oxidase inhibitors on dopamine uptake. Neuropharmacology 33: 763–768

Fozard JR, Zreika M, Robin M, Palfreyman MG (1985) The functional consequences of inhibition of monoamine oxidase type B: comparison of the pharmacological properties of L-deprenyl and MDL 72145. Naunyn Schmiedebergs Arch Pharmacol 334: 186–193

Gelowitz DL, Richardson JS, Wishart TB, Yu PH, Lai C-T (1993) Chronic L-deprenyl or l-amphetamine: equal cognitive enhancement, unequal MAO inhibition. Pharmacol Biochem Behav 47: 41–45

Goldberg SR, Stolerman IP (eds) (1986) Behavioral analysis of drug dependence. Academic Press, London

Goldberg SR, Yasar S, Bergman J (1994) Introduction: examination of clinical and preclinical pharmacologic data relating to abuse liability of l-deprenyl (selegiline). Clin Pharmacol Ther 56: 721–724

Halliday R, Callaway E, Naylar H, Gratzinger P, Prael R (1986) The effects of stimulant drugs on information processing in elderly adults. J Gerontol 41: 748–757

Heikkila RE, Orlansky H, Mytilineou C, Cohen G (1975) Amphetamine: evaluation of d- and l-isomers as releasing agents and uptake inhibitors for ^{3}H-dopamine and ^{3}H-norepinephrine in slices of rat neostriatum and cerebral cortex. J Pharmacol Exp Ther 194: 47–56

Katz JL (1982) Rate-dependent effects of d- and l- amphetamine on schedule-controlled responding in pigeons and squirrel monkeys. Neuropharmacology 21: 235–242

Knoll J, Ecseri Z, Kelemen K, Nievel J, Knoll B (1965) Phenylisopropylmethyl-propinylamine (E-250), a new psychic energizer. Arch Int Pharmacodyn 155: 154–164

Koelega HS (1993) Stimulant drugs and vigilance performance: a review. Psychopharmacology 111: 1–16

Kuczenski R, Segal DS, Cho AK, Melega W (1995) Hippocampus norepinephrine, caudate dopamine and serotonin, and behavioral responses to the stereoisomers of amphetamine and methamphetamine. J Neurosci 15: 1308–1317

Lamb RJ, Griffiths RR (1990) Self-administration in baboons and the discriminative stimulus effects in rats of bupropion, nomifensine, diclofensine and imipramine. Psychopharmacology 102: 183–190

Lynch G, Kessler M, Arai A, Larson J (1990) The nature and causes of hippocampal long-term potentiation. In: Storm-Mathisen J, Zimmer J, Ottersen OP (eds) Progress in brain research, vol 83. Elsevier Science, New York, pp 233–248

Marelt GJ, Vosmer G, Seiden LS (1990) Dopamine uptake inhibitors block long-term neurotoxic effects of methamphetamine upon dopaminergic neurons. Brain Res 513: 274–279

Masand P, Murray GB, Pickett P (1991) Psychostimulants in post-stroke depression. J Neuropsychiatr Clin Neurosci 3: 23–27

Milgram NW, Ivy GO, Head E, Murphy MP, Wu PH, Ruehl WW, Yu PH, Durden DA, Davis BA, Paterson IA, Boulton AA (1993) The effect of L-deprenyl on behavior, cognitive function and biogenic amines in the dog. Neurochem Res 18: 1211–1219

Miller R (1991) Cortico-hippocampal interplay and the representation of contexts in the brain. Springer, Berlin Heidelberg New York Tokyo (Studies of Brain Function, vol 7)

Moser PC (1990) Generalization of L-deprenyl, but not MDL-72974, to the D-amphetamine stimulus in rats. Psychopharmacology 101: S40

Nickel B, Schultze G, Szelényi I (1990) Effect of enantiomers of deprenyl (selegeline) and amphetamine on physical abuse liability and cortical electrical activity in rats. Neuropharmacology 29: 983–992

Philips SR (1981) Amphetamine, p-hydroxyamphetamine and β-phenylethylamine in mouse brain and urine after (−)- and (+)-deprenyl administration. Pharm Pharmacol 33: 739–741

Porsolt RD, Pawelec C, Jalfre M (1984) Discrimination of amphetamine cue: effects of A, B and mixed type inhibitors of monoamine oxidase. Neuropharmacology 23: 569–573

Ramos E, Corsi-Cabrera M, Guevara MA, Arce C (1993) EEG activity during cognitive performance in women. Int J Neurosci 69: 189–195

Reynolds GP, Elsworth JD, Blau K, Sandler M, Lees AJ, Stern GM (1978) Deprenyl is metabolized to methamphetamine and amphetamine in man. Br J Clin Pharmacol 6: 542–544

Risner ME, Jones BE (1977) Characteristics of β-phenylethylamine self-administration by dog. Pharmacol Biochem Behav 6: 689–696

Schechter MD (1978) Stimulus properties of d-amphetamine as compared to l-amphetamine. Eur J Pharmacol 47: 461–464

Shannon HE, De Georgio CM (1982) Self-administration of endogenous trace amines β-phenylethylamine, N-methyl phenylethylamine and phenylethanolamine in dogs. J Pharmacol Exp Ther 222: 52–60

Sprague JE, Nichols DE (1995) The monoamine oxidase-B inhibitor L-deprenyl protects against 3,4-methylenedioxymethamphetamine-induced lipid peroxidation and long-term serotonergic deficits. J Pharmacol Exp Ther 273: 667–673

Spyraki C, Fibiger HC (1981) Intravenous self-administration of nomifensine in rat: implications for abuse potential in humans. Science 212: 11671–168

Tatton WG (1993) "Trophic-like" reduction of nerve cell death by deprenyl without monoamine oxidase inhibition. Neurology Forum 4: 3–10

Taylor KM, Snyder SH (1970) Amphetamine: differentation by d and l isomers of behavior involving brain norepinephrine or dopamine. Science 168: 1487–1489

Terrace HS (1966) Stimulus control. In: Honig WK (ed) Operant behavior: areas of research and application. Prentice-Hall, Englewoods Cliffs NJ, pp 271–344

Timár J, Knoll B (1986) The effect of repeated administration of (−)-deprenyl on the phenylethylamine-induced stereotypy in rats. Arch Int Pharmacodyn 279: 50–60

Warneke L (1990) Psychostimulants in psychiatry. Can J Psychiatry 35: 3–10

Woods SW, Tesar GE, Murray GB, Cassem NH (1986) Psychostimulant treatment of depressive disorders secondary to medical illness. J Clin Psychiatry 47: 12–15

Winger GD, Palmer RK, Woods JH (1989) Drug-reinforced responding: rapid determination of dose-response functions. Drug Alcohol Depend 24: 135–142

Winger GD, Yasar S, Negus SS, Goldberg SR (1994) Intravenous self-administration studies with l-deprenyl (selegiline) in monkeys. Clin Pharmacol Ther 56: 774–780

Wragg RE, Jeste DV (1989) Overview of depression and psychosis in Alzheimer's disease. Am J Psychiatry 146: 577–587

Yasar S, Bergman J (1994) Amphetamine-like effect of l-deprenyl (selegiline) in drug discrimination studies. Clin Pharmacol Ther 56: 768–773

Yasar S, Schindler CW, Thorndike EB, Szelényi I, Goldberg SR (1993a) Evaluation of the stereoisomers of deprenyl for amphetamine-like discriminative stimulus effects in rats. J Pharmacol Exp Ther 265: 1–6

Yasar S, Winger G, Nickel B, Schulze G, Goldberg SR (1993b) Preclinical evalation of l-deprenyl: lack of amphetamine-like abuse potential. In: Szelényi I (ed) Inhibitors of monoamine oxidase B. Birkhäuser, Basel, pp 215–233

Yasar S, Schindler CW, Thorndike EB, Goldberg SR (1994) Evaluation of deprenyl for cocaine-like discriminative stimulus effects in rats. Eur J Pharmacol 259: 243–250

Yokel RA, Pickens R (1973) Self-administration of optical isomers of amphetamine and methylamphetamine by rats. J Pharmacol Exp Ther 187: 27–33

Authors' address: Dr. S. R. Goldberg, NIDA Addiction Research Center, Preclinical Pharmacology Labaoratory, P.O. Box 5180, Baltimore, MD 21224, USA

J Neural Transm (1996) [Suppl] 48: 75–84

Deprenyl in the treatment of Parkinson's disease: clinical effects and speculations on mechanism of action

C. W. Olanow

Mount Sinai School of Medicine, New York, NY, U.S.A.

Summary. Selegiline is a relatively selective inhibitor of monoamine oxidase type B that has been used in Parkinson's disease as an adjunct to levodopa and as putative neuroprotective therapy. Clinical trials demonstrate that selegiline slows the rate of disease progression and delays the appearance of disability necessitating levodopa. However, confounding symptomatic effects have made it difficult to ascertain the presence of any direct neuroprotective effect. Laboratory studies demonstrate that selegiline protects dopaminergic neurons through a mechanism that does not involve MAO-B inhibition. Recent studies suggest that neuroprotection in laboratory models may be related to the capacity of selegiline to up-regulate a series of anti-oxidant and anti-apoptotic molecules which promote cell survival. Further delineation of the precise mechanism whereby selegiline induces this effect may permit for the development of enhanced neuroprotective benefits in PD patients.

Deprenyl (Selegiline), in doses of 5mg BID, is a selective and irreversible inhibitor of monoamine oxidase type B (MAO-B) that is free from the "cheese" effect associated with inhibition of monoamine oxidase type A (MAO-A) in the gut and liver [1,2]. It was originally developed as a possible antidepressant agent in the hope that it could provide benefit without the risk of a sympathomimetic crisis. However, in doses that selectively block MAO-B, deprenyl does not provide significant anti-depressant effects [3,4] and research into the application of this drug as a treatment for depression have been largely abandoned.

Deprenyl was evaluated as a treatment for Parkinson's Disease (PD) based on its capacity to increase striatal dopamine by blocking the degradation of dopamine, promoting its release, and inhibiting its reuptake into dopamine neurons [5]. Birkmayer et al. first tested this hypothesis and reported clinical benefits when deprenyl was added to levodopa in patients with advanced PD [6]. Similar results have been obtained in other studies [7–10] using deprenyl as an adjunct to levodopa. In general, benefits are modest and consist of a reduction in motor fluctuations and an improvement in percent "on" time. These observations form the basis for the approved use of deprenyl in the United States. More recently, there has been interest in the possibility that deprenyl might have a neuroprotective or rescue effect. This notion originally derived from two lines of thinking; the capacity of deprenyl to prevent the development of MPTP-parkinsonism and the possibility that deprenyl might inhibit the development of oxidative stress resulting from the oxidative metabolism of dopamine.

MPTP is a byproduct of the synthesis of an illegal meperidine derivative that induces parkinsonism by way of its oxidation to MPP$^+$ in a reaction catalyzed by MAO-B [11,12].

Inhibition of MAO-B prevents the formation of MPP$^+$ and the development of experimental parkinsonism in animal models [13–15]. To the extent that an environmental toxin such as MPTP contributes to the development of PD, drugs such as deprenyl that inhibit MAO-B activity might protect against the development of neurodegeneration.

The theory that oxidative stress contributes to the pathogenesis of PD has attracted considerable attention because of the potential for the oxidative metabolism of dopamine to generate cytotoxic free radicals [16]. Dopamine is normally metabolised enzymatically (by MAO) or by auto-oxidation to form hydrogen peroxide (H_2O_2). Additionally, iron selectively accumulates withing neuromelanin granules of dopamine neurons [17]. H_2O_2 can react with iron to yield the highly reactive hydroxyl free radical (OH$^.$) according to the Fenton reaction. H_2O_2 is normally cleared by glutathione (GSH), but dysregulation of dopamine metabolism could lead to free radical formation and tissue damage. This could result from increased levels of H_2O_2, increased concentrations of iron, or decreased GSH availability. Indeed, post-mortem laboratory studies provide evidence that the substantia nigra pars compacta (SNc) is in a state of oxidative stress in PD patients [18,19]. Iron, which promotes free radical formation, is increased. Glutathione, which clears H_2O_2, is decreased. And, markers of oxidative damage to a variety of molecules are increased. Further, iron infusion into the SNc of rodents induces a model of parkinsonism secondary to oxidant stress [20]. It can thus be postulated that deprenyl might inhibit H_2O_2 formation derived from MAO-B oxidation of dopamine and limit oxidative damage in PD. Indeed, deprenyl had been shown to prevent oxidative stress induced by an increase in dopamine turnover [21] and is reported to increase longevity in rodents [22]. Further, retrospective studies suggest that PD patients treated with deprenyl plus levodopa liver longer and have less disability than patients treated with levodopa alone [23]. For these reasons, deprenyl was considered to be a rational choice for testing as a possible neuroprotective agent in PD.

Datatop study

Two prospective, randomized, placebo controlled, double-blind trials have been performed to test the hypothesis that deprenyl provides neuroprotective therapy in PD [24–27]. The larger was a multicenter study involving 800 patients known as DATATOP. In these studies, deprenyl monotherapy was compared to placebo in patients with early PD. The primary outcome variable or end point was the time until the emergence of disability sufficient to warrant introduction of levodopa therapy. Both studies showed that deprenyl significantly delayed the emergence of disability necessitating levodopa treatment. However, analyis of the DATATOP study demonstrated that deprenyl is associated with a small but significant symptomatic effect [27] that could mask rather than prevent neuronal degeneration. This effect was not previously known [28] and was not detected in the smaller Tetrud Langston study [24]. Nonetheless, this symptomatic effect confounds interpretation of the mechanism responsible for the effect of deprenyl in these studies and delineation of any putative neuroprotective effect.

Several mechanisms could account for the symptomatic effects detected in deprenyl-treated patients [29]. These include (a) increased striatal dopamine; (b) increased levels of the trace amine phenylethylamine which is a substrate for MAO-B and at high concentrations can exert dopaminergic effects; (c) amphetamine metabolites of deprenyl which can inhibit the uptake and promote the release of dopamine; and (d) an unrecognized deprenyl-related anti-depressant effect. There are, however, reasons to consider that neuroprotection might contribute to the benefits associated with deprenyl treatment. A post hoc analysis of DATATOP patients showed that deprenyl delays the development of disability regardless of whether or not patients experience symptomatic benefits [27]. This does not exclude the possibility that deprenyl acts exclusively by a symptomatic mechanism. It does indicate that if the deprenyl-induced slowing in the emergence of disability in PD is entirely due to the drug's symptomatic effect, this effect does not have to be detected in order to be operative. Further, following drug withdrawal at the end of the study, deprenyl-treated patients had less deterioration from baseline than controls. Accordingly, it is reasonable to consider the possibility that deprenyl might also act through other than a non-symptomatic mechanism.

The sindepar study

To further evaluate the effect of deprenyl on the progression of the signs and symptoms of PD, we performed a prospective, longtitudinal, double-blind, controlled study comparing deprenyl to placebo [30]. This protocol was designed to minimize the drug's symptomatic effects. 101 untreated PD patients were randomly assigned to treatment with either deprenyl or its placebo. In addition, patients were further randomized to receive symptomatic treatment with either levodopa/carbidopa or bromocriptine. Thus, all patients received symptomatic therapy. Patients were withdrawn from deprenyl or its placebo after 12 months and from treatment with levodopa/carbidopa or bromocriptine 7 weeks later. A final visit was performed at month 14, 2 months after withdrawal from deprenyl and 1 week after withdrawal from levodopa/carbidopa or bromocriptine. The primary outcome variable or end point was the change in total UPDRS score between the baseline and final visits. As patients were untreated at both baseline and final visits, the change in UPDRS score between these visits was thought to be an index of disease progression.

Patients randomized to deprenyl had significantly less deterioration in UPDRS score than those on placebo (Fig. 1). Similar effects in favor of

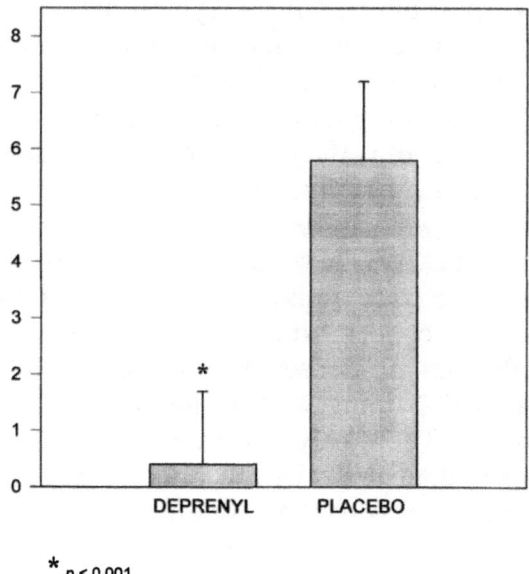

Fig. 1. Histogram demonstrating that deterioration in total UPDRS score ($\pm$ standard error) over 14 months is significantly less in PD patients randomized to deprenyl than those receiving placebo

deprenyl were detected in the sub-group of patients randomized to receive symptomatic treatment with either levodopa/carbidopa or bromocriptine. A significant benefit in favor of deprenyl-treated patients was even detected in a subgroup of 23 patients who underwent a prolonged washout of levodopa/ carbidopa or bromocriptine.

It is possible that the wash-out period for deprenyl was inadequate, as PET scan studies in primates suggest that the half-life for MAO-B regeneration may be 30 days or longer [31]. However, the difference in the extent of deterioration between deprenyl and placebo patients was greater than the full extent of deterioration that occured following withdrawal from Sinemet. If the observed results are interpreted as reflecting inadequate wash-out of deprenyl, this would imply that two month after stopping deprenyl patients will continue to deteriorate by an additional ammount that is greater that the entire deterioration that occurs following withdrawal of Sinemet. Clinical experience indicates that deprenyl has only modest symptomatic effects and that such an explanation is unlikely. Similarly, the results observed in this study are not readily explained by inadequate wash-out of levodopa/ carbidopa or bromocriptine. In this population of patients clinical benefits endure for 7–10 days following withdrawal of levodopa/carbidopa or bromocriptine [32] yet a significant deprenyl effect could still be detected 14 days after withdrawal of these medications despite the small sample size.

Taken together, these findings suggest that a symptomatic mechanism may not fully account for the benefits associated with deprenyl treatment in PD. Critiques of the DATATOP study design do not necessarily apply to the

Sindepar study [33]. Studies showing that deprenyl-treated patients fare no better than those not receiving deprenyl [34,35] have evaluated patients on other symptomatic agents which could mask a neuroprotective benefit and do not preclude the possibility that deprenyl has induced a change in the rate of neuronal degeneration. It is nonetheless evident that any effect that deprenyl has on the natural history of PD is modest. Long-term studies in both early and levodopa-treated advanced PD patients make it clear that deprenyl does not stop disease progression [36,37] and that even if deprenyl induces a degree of neuroprotection, a more effective therapy is required. In this regard, a better understanding of the mechanism(s) responsible for the effects of deprenyl in PD might provide an important clue and permit the development of a more potent neuroprotective therapy.

Mechanism of action

Benefits that have been observed in clinical trials with deprenyl are thought to derive from the capacity of the drug to inhibit MAO-B. This notion is supported by studies with the MAO-B inhibitor Lazabemide which show improvement in PD features comparable to that seen with deprenyl [38,39]. However, it is not clear that MAO-B inhibition fully accounts for the benefits seen with deprenyl in PD. To date, no MPTP-like molecule has been linked to PD. In addition, MPTP-like protoxins capable of causing parkinsonism have been identified that are oxidized by MAO-A [40] and there is no reason at present to think that a putative toxin is more likely to be oxidized by MAO-B than MAO-A. It is also not clear that striatal dopamine is exclusively, or even primarily, oxidized by MAO-B. Dopamine is a substrate for both MAO-A and B, and the relative contribution of these pathways to the metabolism of dopamine in PD is not precisely known. MAO-B has been considered to be the most important because most MAO in the striatum is in the B isoform [41]. However, dopamine metabolism is normally terminated by reuptake into dopamine terminals which primarily contain MAO in the A isoform [42]. It is thus reasonable to speculate that MAO-A may play a significant role in the oxidative metabolism of dopamine. Indeed, following deprenyl administration, cerebrospinal fluid studies homovanillic acid (HVA) levels were only reduced by 20% [43] suggesting that dopamine can be converted to HVA through alternate metabolic pathways. Autooxidation may also play an important role in the metabolism of dopamine. Inhibition of autooxidation in cultured catacholamine-rich neuroblastoma cells is necessary in order to maximally inhibit H_2O_2 formation and protect against levodopa-induced neurotoxicity [44]. It is thus possible that inhibition of MAO-A, MAO-B, and autooxidation might provide superior results to MAO-B inhibition alone if this is the primary mechanism responsible for the effect of deprenyl in PD.

There is also recent laboratory evidence to suggest that deprenyl acts through a mechanism that does not involve MAO-B inhibition. Tatton and colleagues report that mice treated with deprenyl 72 hours after exposure to MPTP experience a significant reduction in neuronal death even though all

MPTP has probably been converted to MPP$^+$ at this time point [45]. Further, these benefits are observed when deprenyl is employed in concentrations that do not inhibit MAO-B. Deprenyl-induced neuronal rescue, using doses that do not inhibit MAO-B, has also been reported by this group in a number of other model systems of neurodegeneration including cultured PC-12 cells deprived of trophic support, survival of retinal ganglion cells, facial axotomy, and ischemia-reperfusion [46]. Others also report studies suggesting that deprenyl has effects that are independent of MAO-B inhibition. Finnegan et al. report that DSP-4 toxicity is blocked by deprenyl through mechanisms independent of MAO-B inhibition [47]. Wu et al., using an in-vivo salicylate trap, observed that dopamine-induced free radical formation and MPP$^+$-induced nigral damage were reduced following deprenyl administration in doses that did not block MAO-B [48]. And, Mytilineou and colleagues noted that deprenyl protects cultured mesencephalic cells against the toxic effect of MPP$^+$ [49], although there is no evidence to implicate MAO-B in the toxicity associated with this compound. While the mechanism by which deprenyl provides benefit in these model systems is not known, there is considerable interest in the possibility that deprenyl acts by inhibiting apoptosis.

Recent evidence indicates that dopaminergic cells can be protected from apoptotic cell death by upregulation of a variety of anti-oxidant and anti-apoptotic proteins [50,51]. A number of neurotoxins including 6 OHDA, levodopa, and dopamine are known to induce cell death by way of apoptosis [52–54]. Apoptosis in these model systems can be inhibited by GSH administration or overexpression of bcl-2. Interestingly, there is now some evidence to suggest that neurotoxins, administered in doses too small to induce apoptosis, may protect dopaminergic cells from other neurotoxins by upregulating cellular defenses. Cultured dopamine neurons exposed to low doses of levodopa show an increase rather than a decrease in GSH and are resistant to butyl-hydrogen peroxide toxicity [55,56]. It is noteworthy that trophic factors such as BDNF similarly induce an increase in GSH and protect cultured dopaminergic cells from neurotoxins [57,58]. Tatton et al. have provided evidence that deprenyl protects injured neurons by inducing upregulation of antioxidant and antiapoptotic proteins and genes such as SOD and

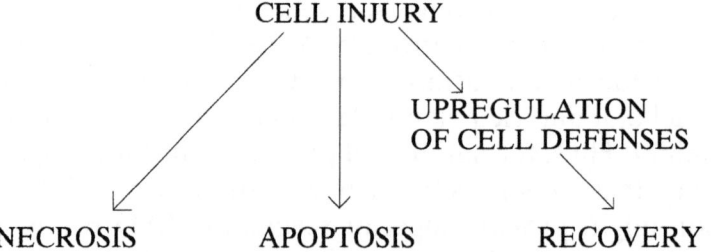

Fig. 2. Schematic representation of hypothesis suggesting that cells exposed to high concentration of a toxin or severe injury die by necrotic death while lesser injury causes death by way of apoptosis. Further, it is hypothesized that a more modest injury or small molecules such as deprenyl may upregulate cellular defenses and protect against apoptosis

BCL-2 [59]. These changes can be blocked by a variety of protein inhibitors suggesting that they are due to new protein synthesis consequent to transcriptional events.

It is interesting to speculate on the possibility that deprenyl renders dopamine neurons relatively resistant to neurodegeneration by upregulating a variety of anti-oxidant and anti-apoptotic defenses (Fig. 2). The finding that deprenyl induces neuronal rescue of a variety of cell types in a variety of model systems is consistent with the notion that its actions are at a cellular level and independent of any specific etiology or pathogenesis. Recent studies in our laboratory support this concept, demonstrating that deprenyl protects against excitotoxic damage through a mechanism that does not involve MAO-B and occurs downstream from the glutamate receptor [60]. Whether these effects rather than MAO-B inhibition account for benefits observed in PD is not established.

Conclusion

The deprenyl story is far from complete but it continues to be a fascinating tale. Deprenyl has been clearly shown to delay the emergence of disability and progression of the signs and symptoms of PD in previously untreated patients. The mechanism responsible for these effects is not yet known nor is it clear if deprenyl has a neuroprotective effect. There is, however, much in the way of clinical and laboratory evidence to suggest that these benefits may relate to other than the drug's small but well-defined symtomatic effects. Of particular interest are the recent findings suggesting that MAO-B inhibition may not prove to be the drug's most important property. The possibility that deprenyl has a more fundamental action that enhances cell defenses and protects them against toxicity and apoptotic cell death is exciting and offers a new direction for investigators hoping to discover a therapy that will provide neuro-protection for PD patients. Further studies will hopefully clarify these issues.

References

1. Knoll J, Esceri Z, Kelemen K, et al (1965) Phenyliopropylmethylpropinylamine (E-250), a new spectrum psychic energizer. Arch Int Pharmacodyn Ther 155: 154–164
2. Elsworth JD, Glover V, Reynolds GP, et al (1978) Deprenyl administration in man: a selective monoamine oxidase B inhibitor without the "cheese effect." Psychopharmacology 57: 33–38
3. Mann JJ, Aarons SF, Wilner PJ, et al (1989) A controlled study of the antidepressant efficacy and side-effects of (−)deprenyl: a selective monoamine oxidase inhibitor. Arch Gen Psychiatry 46: 45–50
4. Mendis N, Paire CMB, Sandler M, et al (1981) Is the failure of (−)deprenyl, a selective monoamine oxidase B inhibitor, to elevate depression related to freedom from the cheese effect? Psychopharmacology (Berlin) 73: 87–90
5. Knoll J (1978) The possible mechanisms of action of (−)deprenyl in Parkinson's disease. J Neural Transm 43: 177–198
6. Birkmayer W, Riederer P, Youdim MBH, Linauer W (1975) The potentiation of the

antiakinetic effect after 1-DOPA treatment by an inhibitor of MAO-B, deprenyl. J Neural Transm 36: 303–326

7. Lees AJ, Kohent LJ, Shaw KM, et al (1977) Deprenyl in Parkinson's disease. Lancet 2: 791–795
8. Eisler T, Teravainen HT, Nelson R, et al (1981) Clinical and biochemical effects of (−)deprenyl in patients with Parkinson's disease: clinical aspects. Neurology 31: 19–23
9. Stern GM, Lees AJ, Hardie RJ, Sandler M (1983) Clinical and pharmacological problems of deprenyl (selegiline) treatment in Parkinson's disease. Acta Neurol Scand 68: 113–116
10. Golbe LI, Lieberman AN, Muenter MD, et al (1988) Deprenyl in the treatment of symptom fluctuations in advanced Parkinson's disease. Clin Neuropharmol 11: 45–55
11. Chiba K, Trevor A, Castagnoli N Jr (1984) Metabolism of the neurotoxic tertiary amine, MPTP, by brain monoamine oxidase. Biochem Biophys Res Comm 120: 457–478
12. Ransom BR, Kunis DM, Irwin I, Langston JW (1987) Astrocytes convert the parkinsonism inducing neurotoxin, MPTP, to its active metabolite, MPP+. Neurosci Lett 75: 323–328
13. Cohen G, Pasik P, Cohen B, et al (1985) Pargyline and deprenyl prevent the neurotoxicity of 1-methyl-4-phenyl-1,2,3,6-tetrahydropyridine (MPTP) in monkeys. Eur J Pharmacol 106: 209–210
14. Langston JW, Irwin I, Langston EB, Forno LS (1984) Pargyline prevents MPTP-induced parkinsonism in primates. Science 225: 1480–1482
15. Heikkila RE, Manzino L, Duvoisin RC, Cabbat FS (1984) Protection against the dopaminergic neurotoxicity of 1-methyl-4-phenyl-1,2,3,6-tetrahydropyridine (MPTP) by monoamine oxidase inhibitors. Nature 311: 467–469
16. Olanow CW (1990) Oxidation reactions in Parkinson's disease. Neurology 40: 32–37
17. Good PF, Olanow CW, Perl DP (1992) Neuromelanin-containing neurons of the substantia nigra accumulate iron and aluminum in Parkinson's disease: a LAMMA study. Brain Res 593: 343–346
18. Jenner P, Schapira AHV, Marsden CD (1992) New insights into the cause of Parkinson's disease. Neurology 42: 2241–2250
19. Olanow CW (1992) Introduction to the free radical hypothesis in Parkinson's disease. Ann Neurol 32: 2–9
20. Sengstock GJ, Olanow CW, Dunn AJ, et al (1992) Iron induces degeneration of substantia nigra neurons. Brain Res Bull 28: 645–649
21. Cohen G, Spina MB (1989) Deprenyl suppresses the oxidant stress associated with increased dopamine turnover. Ann Neurol 26: 689–690
22. Knoll J (1988) Extension of life span of rats by long-term (−)deprenyl treatment. Mt Sinai J Med (NY) 55: 67–74
23. Birkmayer W, Knoll J, Riederer P, et al (1985) Improvement of life expectancy due to 1-deprenyl addition to Madopar treatment in Parkinson's disease: a long-term study. J Neural Transm 64: 113–127
24. Tetrud JW, Langston JW (1989) The effect of deprenyl (selegiline) on the natural history of Parkinson's disease. Science 245: 519–522
25. Parkinson Study Group (1989) DATATOP: a multicenter controlled clinical trial in early Parkinson's disease. Arch Neurol 46: 1052–1060
26. Parkinson's Study Group (1989) Effect of deprenyl on the progression of disability in early Parkinson's disease. NEJM 321: 1364–1371
27. Parkinson's Study Group (1993) Effects of tocopherol and deprenyl on the progression of disability in early Parkinson's disease. NEJM 328: 176–183
28. Yahr MD, Elizan TS, Moros D (1989) Selegiline in the treatment of Parkinson's disease — long term experience. Acta Neurol Scand 128: 157–161
29. Olanow CW, Calne D (1991) Does selegiline monotherapy in Parkinson's disease act by symptomatic or protective mechanisms? Neurology 42: 13–26

30. Olanow CW, Hauser RA, Gauger L, et al (1995) A longtitudinal, double-blind, controlled study of the effect of deprenyl and levodopa on the progression of signs and symptoms of Parkinson's disease. Ann Neurol 38: 771–777
31. Arnett CD, Fowler JS, MacGregor RR, et al (1987) Turnover of brain monoamine oxidase measured in vivo by positron emission tomography using L(11C) deprenyl. J Neurochem 49: 522–527
32. Hauser RA, Olanow CW, Koller WC (1994) Time course of wash-out of symptomatic medication in Parkinson's disease. Neurology 44: 259
33. Schulzer M, Mark E, Calne DB (1992) The antiparkinson efficacy of deprenyl derives from transient improvement that is likely to be symptomatic. Ann Neurol 32: 795–798
34. Brannan T, Yahr MD (1995) Comparative study of selegiline plus L-dopa-carbidopa versus L-dopa-carbidopa alone in the treatment of Parkinson's disease. Ann Neurol 37: 95–98
35. Parkinson Disease Research Group in the United Kingdom (1993) Comparisons of therapeutic effects of levodopa, levodopa and selegiline, and bromocriptine in patients with early, mild Parkinson's disease: three year interim report. BMJ 307: 469–472
36. Elizan TS, Yahr MD, Moros DA, et al (1989) Selegiline use to prevent progression of Parkinson's disease. Experience in 22 de novo patients. Arch Neurol 46: 1275–1279
37. Elizan TS, Yahr MD, Moros DA, et al (1989) Selegiline as an adjunct to conventional levodopa therapy in Parkinson's disease. Experience with this type B monoamine oxidase inhibitor in 200 patients. Arch Neruol 46: 1280–1283
38. Parkinson study group (1993) A controlled clinical trial of lazabemide (RO19-6327) in untreated Parkinson's disease. Ann Neurol 33: 350–356
39. Parkinson Study Group (1994) A controlled trial of Lazabemide (RO19-6327) in levodopa-treated Parkinson's disease. Arch Neurol 51: 342–347
40. Kindt MV, Youngster SK, Sonsalla PK, et al (1988) Role for the monoamine oxidase A (MAO-A) in the bioactivation and nigral striatal dopaminergic neurotoxicity of the MPTP analog 2'ME-MPTP. Eur J Pharmacol 146: 313–318
41. Glover V, Sandler M, Owen F, Riley G (1977) Dopamine is a monoamine oxidase B substrate in man. Nature 265: 80–81
42. Westlund KN, Denney RM, Kochersperger LM, et al (1985) Distinct monoamine oxidase A and B populations in the primate brain. Science 230: 181–183
43. Parkinson's Study Group (1995) Cerebrospinal fluid homovanillic acid in the DATATOP study on Parkinson's disease. Arch Neurol 52: 237–245
44. Pardo B, Mena MA, Fahn S, DeYebenes JG (1993) Ascorbic acid protects against levodopa-induced neurotoxicity on a catecholamine-rich human neuroblastoma cell line. Mov Disord 8: 278–284
45. Tatton WG, Greenwood CE (1991) Rescue of dying neurons: a new action for deprenyl in MPTP parkinsonism. J Neurosci Res 30: 666–677
46. Tatton WG, Ju WYH, Wadia J, Tatton NA (1996) Reduction of neuronal apoptosis by small molecules: promise for new approaches to neurological therapy. In: Olanow CW, Jenner P, Youim MHB (eds) Neurodegeneration and prospects for neuroprotection in Parkinson's disease. Academic Press, London, pp 202–220
47. Finnegan KT, Skratt JJ, Irwin I, et al (1990) Protection against DSP-4 induced neurotoxicity by deprenyl is not related to its inhibition of MAO-B. Eur J Pharmacol 184
48. Wu RM, Chieuh CC, Pert A, Murphy DL (1993) Apparent antioxidant effect of 1-deprenyl on hydroxyl radical formation and nigral injury elicited by MPP^+ in vivo. Eur J Pharmacol 243: 241–247
49. Mytilineou C, Cohen G (1985) Deprenyl protects dopamine neurons from the neurotoxic effect of 1-methyl-4-phenyl-pyridinium ion. J Neurochem 45: 1951–1953
50. Kane DJ, Sarafian TA, Anton R, Hahn H, Gralia EB, Valentine JS, Ord T, Bresdesen DE (1993) Bcl-2 inhibition of neural death: decreased generation of reactive oxygen species. Science 262: 1274–1276

51. Mah SP, Zhong LT, Liu Y, Roghani A, Edwards RH, Bredesen DE (1993) The protooncogene bcl-2 inhibits apoptosis in PC12 cells. J Neurochem 60: 1183–1186
52. Walkinshaw G, Waters CM (1994) Neurotoxin induced cell death in neuronal PC12 cells is mediated by induction of apoptosis. Neuroscience 63: 975–987
53. Walkinshaw G, Waters CM (1995) Induction of apoptosis in catecholaminergic PC12 cells by L-DOPA; implications for the treatment of Parkinson's disease. J Clin Inv 2458–2464
54. Ziv I, Melamed E, Nardi N, Luria D, Achiron A, Offen D, Barzilai A (1994) Dopamine induces apoptosis-like cell death in cultured chick sympathetic. Neurosci Lett 170: 136–140
55. Mytilineou C, Han S-K, Cohen G (1993) Toxic and protective effects of 1-DOPA on mesencephalic cell cultures. J Neurochem 61: 1470–1478
56. Han S-K, Mytilineou C, Cohen G (1996) L-DOPA upregulates glutathione (GSH) and protects mesencephalic cultures against oxidative stress. J Neurochem 66: 501–510
57. Hyman C, Hofer M, Barde YA, et al (1991) BDNF is a neurotrophic factor for dopaminergic neurons of the substantia nigra. Nature 350: 230–233
58. Spina MB, Squinto Sp, Miller J, et al (1991) BDNF protects dopamine neurons against 6-OHDA and MPP$^+$: involvement of the glutathione system. J Neurochem 59: 99–106
59. Tatton WG, Ju WYL, Holland DP, et al (1994) (−)-Deprenyl reduces PC12 cell apoptosis by inducing new protein sythesis. J Neurochem 63: 1572–1575
60. Mytilineou C, Radcliffe P, Leonardi EK, et al (1996) L-Deprenyl protects mesencephalic dopamine neurons from glutamate-receptor-mediated toxicity. J Neurochem (in press)

Authors' address: C. W. Olanow, M.D., F.R. C.P.(C), Mount Sinai School of Medicine, 1 Gustave Levy Place, Annenberg 14-94, New York, NY 10029, U.S.A.

J Neural Transm (1996) [Suppl] 48: 85–93

The clinical potential of Deprenyl in neurologic and psychiatric disorders

W. Kuhn and **Th. Müller**

Department of Neurology, St. Josef-Hospital, Bochum, Federal Republic of Germany

Summary. This article reviews the results of clinical studies with Deprenyl in various neurologic and psychiatric disorders except Parkinson's disease. Promising results could be observed both in narcolepsy in a dose of at least 20 mg/day in three different trials and in one study of Tourette's syndrome including attention hyperactivity disorders using an average dosis of 8.1 mg/day. Controversial results were reported for Alzheimer's disease. On the one hand significant improvement of cognitive functions was found by various authors. On the other hand in a more recent study no effect on the progression of the disease could be observed. For depression a higher dosage of deprenyl between 30 to 60 mg/day appears to be necessary for effective treatment. No positive results were found in amyotrophic lateral sclerosis and in tardive dyskinesias.

Introduction

The mechanism of action of Deprenyl is still under discussion. On the one hand Deprenyl is regarded as selective irreversible inhibitor of cerebral monoamine oxidase type B (MAO-B), resulting in elevated levels of nigrostriatal dopamine (for review see Knoll, 1983; Heinonen and Lammintausta, 1991). However, neuroprotective and neuroregenerative properties of Deprenyl have also been reported in recent years. Increasing experimental data suggest that these effects are presumably not related to inhibition of MAO-B (Tatton, 1993). This promising pharmacological potential has initiated various clinical studies, not only in Parkinson's disease, but also in other neurologic and psychiatric disorders.

1. Neurologic disorders with Parkinsonian features (except Parkinson's disease)

1.1 Progressive supranuclear palsy (PSP)

PSP is a progressive neurodegenerative disease with parkinsonian symptoms like rigidity, bradykinesia, postural instability, dysarthria in combination with supranuclear opthalmoplegia, dysphagia and dementia. A marked depletion of dopamine in the nigrostriatal pathway with selective loss of striatal D_2 — receptors has been demonstrated (Agid et al., 1986). Benefit from drug therapy in PSP is often minimal and side effects can be severe. Therefore a retrospective, uncontrolled case study with 87 patients was performed (Nieforth and Golbe, 1993), to identify drugs with superior adverse event and/ or benefit profiles. The three most frequently used drugs were Amitriptyline, Imipramine and Levodopa/Carbidopa. Levodopa (0.61), Amantadine (0.80), Deprenyl (1.00) and Amitryptiline (1.53) gave the best risk/benefit ratios in contrast to Bromocriptin (2.80) or Desipramine (5.25). 19.5% of patients received Deprenyl with minimal improvement. Minimal adverse effects were observed in 5 patients. Further details regarding dosage and clinical improvement or side effects were not presented. No prospective study with Deprenyl in PSP has been performed up to now.

1.2 Neuroleptic — induced Parkinsonism (NIP)

Anticholinergic drugs are commonly used in the routine treatment of drug-induced parkinsonism. However, other antiparkinsonian substances like Amantadine or even Levodopa have also been reported to improve symptoms of NIP (Baldessarini, 1979; Chouinard et al., 1980). Perenyi et al. (1983) evaluated the effect of Deprenyl on NIP in eleven schizophrenic patients. Deprenyl was administered in daily doses of 2.5 mg b.i.d. on the first six days and 5 mg b.i.d. between the 7th and 21st day of the study. No significant improvement was observed during the treatment in the overall assessment after 3 weeks. However, in 4 patients clinical symptoms improved at least by 50%. No clinical side effects were observed.

A more recent case report was presented by Gewirtz et al. (1993) to test the clinical potential of Deprenyl in longer-term therapy. 10 mg/day of Deprenyl were given to a 54-year-old man with schizophrenia, persistent tremor, rigidity and gait disturbances for at last 7 years that was refractory to standard regimens. Parkinsonism was induced mainly by Perphenazine and Haloperidol. After application of Deprenyl no improvement was noted during the first three months. However, during the fourth month of treatment, a marked improvement of the clinical symptoms about 50% could be observed. No side effects were noted. The mechanism by which Deprenyl improves NIP is not clear. Because increase of dopaminergic tonus occurs very early under Deprenyl, other e.g. regenerative mechanism might be involved.

2. Neurologic disorders without Parkinsonian features

2.1 Amyotrophic lateral sclerosis (ALS)

Based on its neuroprotective and neuroregenerative potential, Deprenyl was tested recently in patients affected by ALS. Mazzini et al. (1994) administered

Deprenyl 10 mg/day orally for 6 months in 53 patients. 58 patients were considered ALS controls. Mortality was similar in the two groups. No statistically significant difference between treated and untreated patients was found both for muscle strength, Norris scale and bulbar score. Similar results were obtained in a double-blind crossover trial with 10 patients suffering from ALS (Jossan et al., 1994). The patients were randomized in such a way that half of the patients started with the active drug and half with the placebo treatment. 10 mg Deprenyl per day were given for 12 weeks and the placebo for the same length of time. A drug free period of 12 weeks between the courses was included. In a preliminary analysis after 36 weeks no statistical significant differences were observed between spinal, Norris and bulbar scores at the two treatment periods. These data are in accordance with preliminary results by Mitchell et al. (1993). Analysis of the first half of this double-blind cross over trial showed no evidence to suggest that Deprenyl might alter the course of ALS over a four month period.

2.2 Narcolepsy

It has been reported that Deprenyl can reduce the total amount of REM sleep and can delay the REM onset in healthy young man (Thornton et al., 1980). For parkinsonian patients, shortened sleep latency was found (Lavie et al., 1980). On the basis of these findings four studies with Deprenyl in narcolepsy have been performed up to now. Schachter et al. (1979) studied 12 patients with narcolepsy and cataplexy. Patients received 5 mg Deprenyl orally twice a day. No detectable therapeutic action could be observed.

Roselaar et al. (1987) examined the effect of 20–30 mg p.o. daily in 21 subjects with narcolepsy over 4 weeks. In this uncontrolled study a reduction of sleep attacks, improvement of subjective alertness, motor coordination, energy and mood was found in treated subjects. It was concluded that Deprenyl requires further evaluation in narcolepsy.

More recently 17 narcolepsy patients were treated in a placebo-controlled, double-blind, crossover trial with 10 to 40 mg daily doses of Deprenyl (Hublin et al., 1994). A dose dependent statistically significant clinical improvement could be observed after 4 weeks. At 40 mg a 36% reduction in the number of daytime sleep episodes and a 34% reduction in their duration was found. The number of cataplectic attacks was reduced by 89%. No intolerable adverse events occured. The effective dose range was 20 to 40 mg. However, low-tyramine diet was necessary.

These data are in accordance with the results of a recently published study (Mayer et al., 1995). In a double-blind, placebo-controlled design patients were randomly assigned to three groups (placebo, 2×5 mg and 2×10 mg Deprenyl). After two weeks daytime sleepness improved significantly. The number of sleep attacks and naps as well as the frequency of cataplexy were reduced. It was concluded that Deprenyl at a dose of at least 20 mg/day is a potent drug for the treatment of narcoleptic syndroms.

2.3 Other neurologic disorders

Deprenyl has also been tested for efficacy in the prophylaxis of migraine without aura. In an open study no significant improvement has been reported (Kuritzky et al., 1992).

In the case of postpolio syndrome two patients have been treated with Deprenyl. The first patient developed mild Parkinsonian symptoms, for which he was treated with Levodopa/Carbidopa followed by addition of Deprenyl. An improvement in the symptoms of postpolio syndrome could be observed similar to a patient who was unaffected by Parkinson's disease and was on Deprenyl alone (Bamford et al., 1993).

3. Neuropsychiatric disorders

3.1 Alzheimer's disease (AD)

The rationale for the use of Deprenyl in AD is based on data from post mortem studies. MAO-B levels in the brain have been shown to be elevated in AD (Oreland and Gottfries, 1986). Further, Nakamura et al. (1990) found that astrocytes associated with Alzheimic plaques express exclusively MAO-B.

The first clinical study was performed by Tariot et al. (1987a,b), who administered 10 mg and 40 mg/day of Deprenyl to 17 patients with AD in a double-blind, placebo-controlled serial treatment. They observed improvement of anxiety, depression, tension, episodic memory, learning and a decrease of excitement. One half of the patients showed increased activity and social interaction. Similar but smaller changes were observed during 40 mg/day treatment.

Long-term effects have also been demonstrated. Under double-blind conditions significant improvement both in memory, attention (Piccinin et al., 1990; Agnoli et al., 1990) and activities of daily living, orientation, word fluency and visual spatial ability (Mangnoni et al., 1991) after 3 month of treatment with 10 mg/day of Deprenyl have been reported.

These results confirm other positive results obtained with trials of smaller sample size (Campi et al., 1990; Monteverdi et al., 1990). More recently 5 mg b.i.d. of Deprenyl were added in a double-blind cross over study to 10 AD patients receiving either Tacrine or Physostigmine. It was found that Deprenyl was associated with significant improvement in cognition (Schneider et al., 1993). These findings are in contrast to the results of a 15-month randomized, double-blind placebo-controlled trial reported by Burke et al. (1993). The total score of the Brief Psychiatric Rating Scale was significantly less after 15 months in 39 subjects taking Deprenyl. However, no improvement of behavior or cognitive function could be observed. It was concluded that Deprenyl did not appear to slow the progression of the disease.

3.2 Tardive Dyskinesia (TD)

It has been suggested that oxidative stress may be a primary factor in the etiology of TD (Lohr, 1991). Neuroleptic treatment may increase turnover of dopamine and thus increase formation of cytotoxic radicals. Deprenyl has been reported to lower levels of glutathione disulfide in rats treated with haloperidol, indicating a reduction in oxidative stress (Cohen and Spina, 1989).

To explore further the role of antioxidant agents for the treatment of TD, a double-blind, placebo-controlled trial over 6 weeks was performed (Goff et al., 1993). 33 patients with TD were randomly assigned to Deprenyl 10 mg/day or placebo. 28 subjects completed at least 1 week of treatment. The group receiving Deprenyl displayed significantly less improvement of TD compared with the placebo group. Therefore Deprenyl was less effective than placebo in reducing symptoms of TD. It was speculated that this may be the result of dopamine agonist effects associated with Deprenyl (Goff et al., 1993).

3.3 Tourette's Syndrome (TS)

TS is a neurologic disorder characterized by involuntary motor and phonic tics and various behaviorial disturbances including attention hyperactivity disorders (ADHD; Jankovic, 1992). Because Deprenyl may have a stimulatory effect and MAO inhibitors have been shown to ameliorate hyperactive behavior, a study was conducted in children with the Tourette's syndrome — ADHD combination (Jankovic, 1993). 29 patients were enrolled in this open trial. The average duration daily dose was 8.1 mg/day. 26 of all patients (90%) reported clinically meaningful improvement in their ADHD. No serious side effects could be observed. Two patients noted exacerbation of their tics. These results suggest that Deprenyl can be a safe and effective treatment of ADHD in patients with Tourette's syndrome.

4. Psychiatric disorders

4.1 Schizophrenia

It has been speculated on the basis of earlier studies that MAO-A inhibitors may be beneficial for the relief of negative symptoms in schizophrenia (Crow, 1980). Bucci (1987) found the nonselective MAO inhibitor tranylcypromine to be effective in the treatment of negative symptoms. Negative symptoms have been related to dopamine hypoactivity (Meltzer et al., 1986). Because of the stimulating effect of Deprenyl on the dopaminergic system, an open trial has been performed with 13 male chronic schizophrenic patients (Perenyi et al., 1992). None of the patients was suffering from clinically relevant tardive dyskinesia. Deprenyl was added to the ongoing antipsychotic therapy. 5–

10 mg were applied to day ten. Between day 11 and 45 the dose was 15 mg/day. Significantly lower total scores were reported on the Negative Symptoms Rating Scale (BPRS). No significant changes in the BPRS total score related to positive symptoms could be observed. It was concluded that the risk of induction of positive symptoms is very low when Deprenyl is used under the condition of this study (Perenyi et al., 1992).

4.2 Depression

The use of MAO inhibitors for the treatment of Depression has been limited by the tyramine-induced hypertensive crisis ("cheese effect"), which is caused by inhibition of gut MAO-A. The observation that Deprenyl selectively inhibits MAO-B (Knoll and Magyar, 1972) has initiated various clinical studies in depression. Early clinical trials showed Deprenyl to be an effective antidepressant (Varga and Tringer, 1967; Tringer, 1971). This has been only partially confirmed by other authors (Mann et al., 1982; Quitkin et al., 1982). From these studies it has been concluded that Deprenyl may have antidepressant properties especially in atypical, nonendogenous and bipolar depression. However, only few double-blind placebo-controlled trials have been performed up to now.

Mendis et al. (1981) treated 11 patients with 20 mg over 3 weeks. No significant effect on unipolar endogenous depression could be observed. In contrast, 14 affectively ill patients (2 bipolar, 12 unipolar) shared significantly greater clinical improvement than placebo patients under 15 mg Deprenyl (Mendlewicz and Youdim, 1983). Mann et al. (1989) performed a double-blind placebo-controlled study with 10 mg in comparison to about 30 mg/day in 22 depressed outpatients. The lower dose did not have a statistically significant antidepressant effect after 3 weeks. However, after 6 weeks and at higher doses Deprenyl was superior to placebo without cheese effect. It was concluded that Deprenyl is an effective antidepressant in dosages between 10 to 50 mg/day. The possible importance of high-dose Deprenyl has been confirmed in a recent study, where 16 treatment-resistant older depressive patients entered a double-blind, randomized crossover study of placebo vs. 3 weeks of Deprenyl at a dosage of 60 mg/day (Sunderland et al., 1994). A significant improvement of mood and behavior could be observed. However all subjects were maintained on low tyramine diet. These results suggest that Deprenyl appears to be an effective antidepressant in dosages between 30–60 mg/day, probably due to partial inhibition of MAO-A.

References

Agid Y, Javoy-Agid F, Ruberg M, et al (1986) Progressive supranuclear palsy: anatomo-clinical and biochemical considerations. Adv Neurol 45: 191–206

Agnoli A, Martucci N, Fabbrini G, et al (1990) Monoamine oxidase and dementia: treatment with an inhibitor of MAO-B activity. Dementia 1: 109–114

Baldessarini RJ (1979) Neurological toxicology of antipyschotic drugs. McLean Hosp J 4: 2–19

Bamford CR, Montgomery EB jr, Munoz JE, et al (1993) Postpolio syndrome-response to deprenyl (selegiline). Int J Neurosci 71(1–4): 183–188

Bucci L (1987) The negative symptoms of schizophrenia and the monoamine oxidase inhibitors. Psychopharmacology 91: 104–108

Burke WJ, Roccaforte WH, Wengel SP, et al (1993) L-Deprenyl in the treatment of mild dementia of the Alzheimer type: results of a 15-month trial. J Am Geriatr Soc 41(11): 1219–225

Campi N, Todeschini GP, Scarzella L (1990) Selegiline versus L-acetylcarnitine in the treatment of Alzheimer type dementia. Clin Ther 12: 306–314

Chouinard G, Jones BD, Annable L (1980) L-dopa in neuroleptic induced extrapyramidal symptoms. 35th Annual Convention of Society of Biological Psychiatry of USA, Boston

Cohen G, Spina MB (1989) Deprenyl suppresses the oxidant stress associated with increased dopamine turn-over. Ann Neurol 26: 689–690

Crow TJ (1980) Molecular pathology of schizophrenia: more than one disease process? Br Med J 280: 66–68

Gewirtz GR, Sharif Z, Cadet JL, Sarti P, Gorman JM (1993) Selegiline for neuroleptic-induced Parkinsonism. Pharmacopsychiat 26: 128–129

Goff DC, Renshaw PF, Sarid-Segal O, et al (1993) A placebo-controlled trial of selegiline (L-deprenyl) in the treatment of tardive dyskinesia. Biol Psychiatry 33: 700–706

Heinonen EH, Lammintausta R (1991) A review of the pharmacology of selegiline. Acta Neurol Scand 136 [Suppl]: 44–59

Hublin C, Partinen M, Heinonen EH, Punkka P, Salmi T (1994) Selegiline in the treatment of narcolepsy. Neurology 44: 2095–2101

Jankovic J (1992) Diagnosis and classification of tics and Tourette Syndrome. Adv Neurol 58: 7–14

Jankovic J (1993) Deprenyl in attention deficit associated with Tourette's syndrome. Arch Neurol 50: 286–288

Jossan SS, Ekblom J, Gudjonsson O, Hagbarth KE (1994) Double blind cross over trial with deprenyl in amyotrophic lateral sclerosis. J Neural Transm [Suppl] 41: 237–241

Knoll J (1989) The pharmacology of selegiline ((−)deprenyl). New aspects. Acta Neurol Scand 126 [Suppl]: 83–91

Knoll J, Magyar K (1972) Some puzzling pharmacological effects of monoamine oxidase inhibitors. In: Costa E, Sandler M (eds) Monoamine oxidase — new vistas. Raven Press, New York, pp 393–408

Kuritzky A, Zoldan Y, Melamed E (1992) Selegiline, a MAO-B inhibitor, is not effective in the prophylaxis of migraine without aura — an open study. Headache 32(8): 416

Lavie P, Wajsbort J, Youdim MBH (1980) Deprenyl does not cause insomnia in parkinsonian patients. Commun Psychopharmacol 4: 303–307

Lohr JB (1991) Oxygen radicals and neuropsychiatric illness. Arch Gen Pschiatry 48: 1097–1106

Mangoni A, Grassi MP, Frattola L, Piolti R, Bassi S, Motta A, Marcone A, Smirne S (1991) Effects of a MAO-B inhibitor in the treatment of Alzheimer's disease. Eur Neurol 31: 100–107

Mann J, Gershon S (1980) L-Deprenyl, a selective monoamine oxidase type-B inhibitor in endogenous depression. Life Sci 26: 877–882

Mann JJ, Frances A, Kaplan RD, et al (1982) The relative efficacy of 1-Deprenyl, a selective monoamine oxidase type B inhibitor, in endogenous and nonendogenous depression. J Clin Psychopharmacol 2(1): 54–57

Mann JJ, Aarons SF, Wilner PJ, Keilp JG, Sweeney JA, Pearlstein T, Frances AJ, Kocsis JH, Brown RP (1989) A controlled study of the antidepressant efficacy and side effects of (−)-deprenyl: a selective monoamine oxidase inhibitor. Arch Gen Psychiatry 46: 45–50

Mayer G, Ewert Meier K, Klinik Hephata (1995) Selegiline Hydrochloride treatment in narcolepsy. A double-blind, placebo-controlled study. Clin Neuropharmacol 18/4: 306–319

Mazzini L, Testa D, Balzarini C, Mora G (1994) An open-randomized clinical trial of selegiline in amyotrophic lateral sclerosis. J Neurol 241: 223–227

Meltzer HY, Sommers AA, Luchins DJ (1986) The effect of neuroleptics and other psychotropic drugs on negative symptoms in schizphrenia. J Clin Psychopharmacol 6: 329–338

Mendis N, Pare CMB, Sandler M, et al (1981) Deprenyl in the treatment of depression. In: Youdim MBH, Paykel ES (eds) Monoamine oxidase inhibitors: the state of the art. John Wiley and Sons, Chichester, pp 171–176

Mendlewicz J, Youdim MBH (1983) L-Deprenyl, a selective monoamine oxidase type B inhibitor, and the treatment of depression: a double blind evaluation. Br J Psychiatry 142: 508–511

Mitchell JD, Houghton E, Kilshow J, et al (1993) Free radicals, sporadic motor neurone disease and selegiline. Symposium on Neuroprotection and clinical trials in MND/ ALS, Paris (Abstract)

Monteverde A, Gnemmi P, Rossi F, Monteverde A (1990) Selegiline in the treatment of mild to moderate Alzheimer type dementia. Clin Ther 12: 315–322

Nakamura S, Kawamata T, Akiguchi I, Kameyama M, Nakamura N, Kimura H (1990) Expression of monoamine oxidase B activity in astrocytes of senile plaques. Acta Neuropathol 80: 419–425

Nieforth KA, Golbe LI (1993) Retrospective study of drug response in 87 patients with progressive supranuclear palsy. Clin Neuropharmacol 16/4: 338–346

Oreland L, Gottfries CG (1986) Brain and brain monoamine oxidase in aging and in dementia of Alzheimer's type. Prog Neuropsychopharmacol Biol Psychiatry 10: 533– 540

Perenyi A, Bagdy G, Arato M (1983) An early phase II trial with l-deprenyl for the treatment of neuroleptic-induced Parkinsonism. Pharmacopsychiat 16: 143–146

Perenyi A, Goswami U, Frecska E, Arató M, Bela A (1992) L-Deprenyl in treating negative symptoms of schizophrenia. Psychiatry Res 42: 189–191

Piccinin GL, Finali G, Piccarilli M (1990) Neuropsychological effects of L-deprenyl in Alzheimer's dementia. Clin Neuropharmacol 13: 147–163

Quitkin FM, Liebowitz MR, Stewart JW, McGrath PJ, Harrison W, Rabkin JG, Markowitz J, Davies SO (1984) L-Deprenyl in atypical depressives. Arch Gen Psychiatry 41: 777–781

Quitkin FM, Liebowitz MR, Stewart JW, et al (1985) Deprenyl in atypical depressives. I. Clinical efficacy. Proceedings of the International Symposium in Deprenyl. Chinoin, Budapest, pp 115–119

Roselar SE, Langdon N, Lock CB, Jenner P, Parkes JD (1987) Selegiline in narcolepsy. Sleep 10(5): 491–495

Schachter M, Price PA, Parkes JD (1979) Deprenyl in narcolepsy. Lancet i: 831–832

Schneider LS, Olin JT, Pawluczyk S (1993) A double-blind crossover pilot study of l-deprenyl (selegiline) combined with cholinesterase inhibitor in Alzheimer's disease. Am J Psychiatry 150: 321–323

Sunderland T, Cohen R, Molchan S, et al (1994) High dose selegiline in treatment-resistant older depressive patients. Arch Gen Psychiatry 51: 607–615

Tariot PN, Cohen R, Sunderland T, et al (1987a) L-Deprenyl in Alzheimer's disease. Arch Gen Psychiatry 44: 427–433

Tariot PN, Sunderland T, Weingartner H, et al (1987b) Cognitive effects of L-deprenyl in Alzheimer's disease. Psychopharmcology 91: 489–495

Tatton WG (1993) Selegiline can mediate neuronal rescue rather than neuronal protection. Mov Disord 8: 520–530

Thornton C, Dore CJ, Elsworth JD, Herbert M, Stern GM (1980) The effect of Deprenyl, a selective monoamine oxidase B inhibitor, on sleep and mood in man. Psychopharmcol 70: 163–166

Tringer L, Haits G, Varga E (1971) The effect of L-E-280 (L-phenylisopropyl-methyl-propinylamine) in depression. Soc Pharmacol Hungarica, Budapest, pp 111–113

Varga E, Tringer L (1967) Clinical trial of a new type of promptly acting psychoenergetic agent (Phenyl-isopropyl-methyl propinyl-HCL, E 250). Acta Med Acad Sci Hung 23: 289–295

Authors' address: Prof. Dr. W. Kuhn, Neurologische Universitätsklinik Bochum, St. Josef-Hospital, Gudrunstrasse 56, D-44791 Bochum, Federal Republic of Germany

J Neural Transm (1996) [Suppl] 48: 95–101

Pharmacology and neuroprotective properties of rasagiline

J. P. M. Finberg[1], **I. Lamensdorf**[1], **J. W. Commissiong**[2], and
M. B. H. Youdim[1,3]

[1] Pharmacology Unit, Rappaport Faculty of Medicine, Technion, Haifa, Israel
[2] NINDS, NIH, Bethesda, MD, USA
[3] Rappaport Family Research Institute, Haifa, Israel

Summary. Rasagiline [R(+)-N-propargyl-1-aminoindane] is a selective irreversible inhibitor of MAO-B which is not metabolised to amphetamine-like derivatives. Like deprenyl, when given to rats in a dose selective for inhibition of MAO-B, it does not affect striatal extracellular fluid dopamine levels, but when administered chronically (21 days) it increased striatal microdialysate dopamine without reduction in deaminated metabolites. Similarly to deprenyl, rasagiline (10^{-6}M) increased the percentage of tyrosine hydroxylase positive cells in a primary culture of rat fetal mesencephalic cells (6 days in culture). Rasagiline, but not deprenyl, also increased the number of neurons per field in this organotypic culture.

Introduction

The success of (−)-deprenyl as an adjunct to L-DOPA treatment of Parkinson's disease (Birkmayer et al., 1975) has prompted the development of other selective inhibitors of MAO-B. The antiparkinsonian effect of deprenyl is thought to comprise both symptomatic and neuroprotective components, although the extent of each of these elements in attaining clinical antiparkinsonian action is unclear. The former may depend on an increase in synaptic dopamine (DA) and phenylethylamine when deprenyl is administered alone or together with L-DOPA, and both inhibition of MAO, and amphetamine-like properties of the molecule may play a part in achieving this effect (Knoll, 1986; Riederer and Youdim, 1986; Riederer et al., 1986; Tatton, 1993). The latter has been demonstrated definitively only in isolated cell or intact animal experiments (Tatton et al., 1994; Tatton and Greenwood, 1991).

We have previously described an indane derivative, AGN-1135 (N-propargyl-1-aminoindane; Kalir et al., 1981; Finberg et al., 1981a, b), which produced selective inhibition of MAO-B without the amphetamine-like properties of deprenyl. Since AGN-1135, as deprenyl, is a chiral compound, we decided to investigate the activity of its optical isomers, and discovered that one isomer, R(+)-N-propargyl-1-aminoindane (rasagiline), possesses nearly all of the enzyme-inhibitory activity. We have therefore further investigated effects of this isomer on striatal DA release by microdialysis, and on isolated dopaminergic cell survival in culture. Rasagiline has been used in the form of the mesylate (TVP-1012) or hydrochloride (TVP-101) salts.

Materials and methods

MAO activity

Inhibition of MAO activity was studied in vitro and ex vivo. Enzyme activity was measured in sucrose homogenates of tissues by incubation with ^{14}C-labelled 2-phenylethylamine (MAO-B) or 5-hydroxytryptamine (MAO-A). Labelled metabolites were separated by ion exchange chromatography, or by solvent extraction, using a modification of the general methods of Otsaka and Kobayashi (1964) and O'Carroll et al. (1983).

Striatal microdialysis

Two series of experiments were performed. In the first (Finberg et al., 1995), microdialysis was performed under halothane anesthesia, and rasagiline was administered 2 h prior to collection of microdialysate. Striatal DA release was measured in basal condition and following local striatal infusion of L-DOPA. In the second series of experiments, rasagiline or deprenyl were administered either once only, or daily for 21 days, by s.c. injection, and microdialysis was performed 24 h after last injection. Microdialysis probes (4 mm active length) were implanted in rat striatum under pentobarbital/chloral hydrate anesthesia (12/60 mg/kg). Striatal DA release was studied the following day, 24 h after last dose of drug, in basal condition, and following local striatal infusion of KCl (60 mM) via microdialysis probe.

Concentration of DA and metabolites (DOPAC, HVA) in microdialysate was determined by HPLC/EC. Electrochemical detection of amines and metabolites was carried out using a Waters amperometric detector for first series, and an ESA Coulochem detector for second series.

Effects of rasagiline on rat fetal dopaminergic cells in culture

Rat fetal mesencephalic cells from 14 day embryos were prepared as described by Shimoda et al. (1992). These cultures contain 95% neurons, of which about 20% are tyrosine hydroxylase positive when measured between 4 h and 7 days after plating. The cells were grown in chamber slides precoated with poly-D-lysine. Cell viability was determined using a two color fluorescence viability assay (Molecular Probes), or with Neuro Tag. Number of cells per well was determined by counting all of the cells in a chamber using a low-power eyepiece, in order to allow for uneven distribution of the cells. Drugs were added to the medium at time of plating, and were present at the stated concentrations up to time of counting, at day 6. The effect of MAO inhibitors was studied both in cells cultured in serum-enriched medium, and also in the absence of serum. The latter experiments were carried out using the medium described by Takeshima et al. (1994). Presence of tyrosine hydroxylase in cells was determined by the immunochemical technique described by Shimoda et al. (1992).

Results

MAO inhibition

When rat brain mitochondria were incubated in the presence of rasagiline for 20 min before addition of labelled substrates, a selective MAO-B inhibitory effect was apparent (IC_{50} values 2.3 and 149 nM for MAO-B and MAO-A respectively), whereas the S(−) isomer of rasagiline was about 3 orders of magnitude less potent. Following oral administration to rats, ED_{50} values (mg/kg) for inhibition of brain MAO-B and MAO-A by rasagiline were 0.1 and 6 respectively. The degree of selectivity for MAO-B compared to MAO-A inhibition is similar for rasagiline and deprenyl, but rasagiline is about 5 times more potent as an MAO-B inhibitor than deprenyl following single po administration in the rat.

Striatal microdialysis

In the halothane anesthetised rat, acute treatment with clorgyline increased striatal efflux of DA in response to local striatal infusion of L-DOPA (1 mM for 10 min via the microdialysis probe), whereas acute treatment with deprenyl or rasagiline had no effect on striatal DA efflux (Finberg et al., 1995). In this series of experiments, sensitivity of DA detection was not adequate to accurately measure basal striatal microdialysate level.

In conscious rats, resting levels of striatal DA were well within detection limit. Neither deprenyl nor rasagiline affected striatal DA microdialysate level following acute administration, but after 3 weeks daily injection of rasagiline (0.05 mg/kg s.c.) or deprenyl (0.25 mg/kg s.c.), basal microdialysate level was significantly increased (276 and 260%, $P < 0.05$). Striatal DA level following local administration of KCl was also significantly greater in rats treated chronically with deprenyl or rasagiline. The doses of clorgyline, deprenyl and rasagiline used in this study caused selective inhibition of MAO-A (clorgyline) or MAO-B (deprenyl, rasagiline) with less than 30% inhibition of the opposite form. These data, together with MAO-inhibitory properties of rasagiline, will be published in full elsewhere.

Dopaminergic cells in culture

Both deprenyl and rasagiline caused a marked stimulation of neurite outgrowth, as described by Roy and Bedard (1993). When cultured in serum-containing medium, both drugs caused a significant increase in the percentage of tyrosine hydroxylase positive neurons at a concentration of 1.0×10^{-6} M, but only rasagiline caused an increase in number of neurons per field at 10^{-7} and 10^{-6} M (Fig. 1). Clorgyline (1.0×10^{-6} M and 1.0×10^{-7} M) had no effect on cell survival or on percentage of tyrosine-positive cells. When the cells

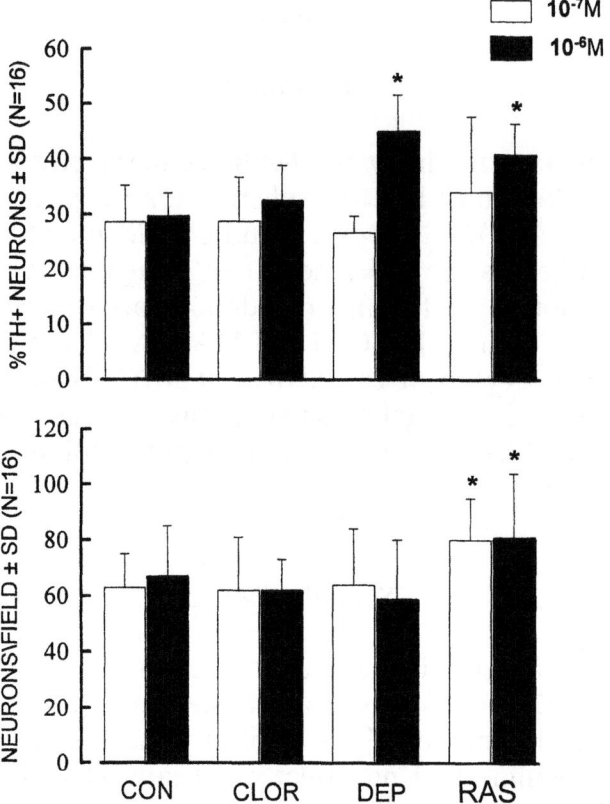

Fig. 1. Effect of MAO inhibitors on survival of rat fetal mesencephalic neurons in culture. Number of neurons per field, and percentage tyrosine hydroxylase positive neurons, were counted following 6 days in culture in medium containing clorgyline (CLOR), deprenyl (DEP), rasagiline (RAS) or diluent (CON) at 10^{-7} (open columns) or 10^{-6} (closed columns) molar. Mean values shown $\pm$ SD for n = 16 wells at each concentration.* P < 0.05

were cultured in the absence of serum, the presence of rasagiline (1.0 $\times$ 10^{-6}M) caused an increase in the number of tyrosine hydroxylase positive neurons (from 2.5 to 5.2%, P < 0.05), but none of the MAO inhibitors increased number of cells surviving at time of counting.

Discussion

In previous studies, we have shown that the racemic form of rasagiline (AGN-1135) is devoid of amphetamine-like effects on cardiovascular system, and in the cat nictitating membrane, whereas deprenyl exerts a distinct sympathomimetic effect in these preparations (Simpson, 1978; Finberg et al., 1981b; Finberg and Youdim, 1985). In separate studies, we have confirmed the absence of amphetamine-like activity of rasagiline in the pithed rat at doses up to 10 mg/kg iv. Rasagiline is well absorbed after oral administration, and efficiently passes the blood-brain barrier, as shown by the high degree of

MAO inhibition in rat brain tissue. Preclinical toxicity testing has shown absence of untoward toxic effects in laboratory animals, and this has been confirmed in phase 1 clinical trials in human volunteers. The high degree of potency for MAO-B inhibition has been confirmed in humans (1 mg daily causes more than 90% inhibition of platelet MAO-B after 7 days).

Clinical studies with deprenyl in PD patients have shown that the drug causes symptomatic improvement when given in monotherapy, as well as enhancing the effect of L-DOPA. The mechanism of this effect is currently unclear, since dopaminergic nerves of the nigro-striatal tract contain MAO-A in rodents and man (Westlund et al., 1985). Extra-dopaminergic nerve deamination of DA in astrocytes expressing MAO-B could explain this property of deprenyl. In the rat, microdialysis experiments have confirmed observations on tissue amine levels, showing that DA behaves as a substrate of MAO-A (Kato et al., 1986; Buu and Angers, 1987; Butcher et al., 1990; Colzi et al., 1990). Our microdialysis studies, both in anesthetised and conscious rats, have also shown that following a single injection of inhibitor, striatal microdialysate DA is affected by pretreatment with clorgyline but not deprenyl or rasagiline, when the inhibitors are administered at MAO-A or MAO-B selective doses. Following chronic administration of deprenyl or rasagiline, however, striatal DA release was enhanced. The mechanism of this effect is currently under study. It could be caused by an inhibition of neuronal or astrocytic DA uptake which develops slowly with time, or it could be explained by accumulation of endogenous β-phenylethylamine, as proposed by Paterson et al. (1991). Chronic administration of β-phenylethylamine causes a gradually developing potentiation of dopamine release from the striatum (Kuroki et al., 1990). Other studies have found a similar action of chronically as opposed to acutely administered deprenyl to enhance release of DA from striatal tissue in vitro (Knoll, 1986; Tekes et al., 1988). Certain experimental evidence with deprenyl have been interpreted as ruling out a role of MAO inhibition or uptake inhibition in producing this effect (Knoll, 1992). Our findings with rasagiline, a rigid structure non-amphetamine-like propargyl inhibitor, indicate that this ability of deprenyl to enhance striatal DA release on chronic administration is not dependent on an amphetamine-like action of the inhibitor.

In addition to the dopaminergic effects of deprenyl and rasagiline, both inhibitors possess neuroprotective and neurorescue properties. Deprenyl has been shown to prevent cell death in newborn rat facial neurons deprived of trophic support at doses below those required for MAO inhibition (Tatton and Greenwood, 1991; Salo and Tatton, 1992; Tatton, 1993). MAO-B inhibition by other selective inhibitors such as MDL-72974A (irreversible) or RO-16-6491 (reversible) does not prevent cell death in serum-deprived PC-12 cells, whereas deprenyl and pargyline do (Tatton et al., 1994). Our observations in rat fetal mesencephalic cells in culture show that rasagiline also possesses neuroprotective properties, as seen by reduction of the rate of spontaneous cell death seen in cell culture in vitro, and reduction in accelerated cell death following serum deprivation. It should be noted that in the serum deprivation model used in our study, the dopaminergic cells have not been exposed at all to serum. In these initial results, rasagiline showed the ability both to increase numbers of surviving cells, as well as the survival of tyrosine hydroxylase positive cells, whereas deprenyl increased the proportion of tyrosine hydroxylase positive cells only. This effect of deprenyl was also shown by Roy and Bedard (1993). Currently it is not clear whether deprenyl and rasagiline selectively increase survival of cells which express tyrosine hydroxylase, or increase expression of the enzyme within surviving cells.

The mechanism of the neuroprotective and neurorescue properties of rasagiline, as with deprenyl, is currently not understood (see elsewhere in this symposium). Both compounds seem to possess neurotrophic properties. Serum deprivation in PC-12 cells leads to apoptotic cell death, and recent results indicate a similar process for our primary dopaminergic cells cultured in the absence of serum (JW Commissiong, personal communication). The effect of deprenyl to increase cell survival in PC-12 cells is dependent on synthesis of new protein, and is therefore not a simple anti-oxidant property of the drug (Tatton et al., 1994).

These and other observations on rasagiline in this symposium indicate that rasagiline may be an advance over deprenyl in treatment of Parkinson's and other neurodegenerative diseases, in possessing neuroprotective and neurorescue properties uncomplicated by production of potentially neurotoxic metabolites such as (−)amphetamine and (−)methamphetamine.

References

Birkmayer W, Riederer P, Youdim MBH, Linnauer W (1975) The potentiation of the antiakinetic effect after L-dopa treatment by an inhibitor of MAO-B deprenil. J Neural Transm 36: 303–326

Butcher SP, Fairbrother IS, Kelly JS, Arbuthnott GW (1990) Effects of selective monoamine oxidase inhibitors on the in vivo release and metabolism of dopamine in the rat striatum. J Neurochem 55: 981–988

Buu NT, Angers M (1987) Effects of different monoamine oxidase inhibitors on the metabolism of L-dopa in the rat brain. Biochem Pharmacol 36: 1731–1735

Colzi A, d'Agostini F, Kettler R, Borroni E, Da Prada M (1990) Effect of selective and reversible MAO inhibitors on dopamine outflow in rat striatum: a microdialysis study. J Neural Transm [Suppl] 32: 79–84

Finberg JPM, Youdim MBH (1985) Modification of blood pressure and nictitating membrane response to sympathomimetic amines by selective monoamine oxidase type A and type B inhibitors. Br J Pharmacol 85: 541–546

Finberg JPM, Tenne M, Youdim MBH (1981a) Tyramine antagonistic properties of AGN 1135 — an irreversible inhibitor of monoamine oxidase type B. Br J Pharmacol 73: 65–74

Finberg JPM, Tenne M, Youdim MBH (1981b) Selective irreversible propargyl derivative inhibitors of monoamine oxidase (MAO) without the cheese effect. In: Youdim MBH, Paykel ES (eds) Monoamine oxidase inhibitors — the state of the art. Wiley, Chichester, pp 31–43

Finberg JPM, Wang J, Goldstein DS, Kopin IJ, Bankiewicz KS (1995) Influence of selective inhibition of monoamine oxidase A or B on striatal metabolism of L-DOPA in hemiparkinsonian rats. J Neurochem 65: 1213–1220

Kalir A, Sabbagh A, Youdim MBH (1981) Selective acetylenic "suicide" and reversible inhibitors of monoamine oxidase type A and B. Br J Pharmacol 73: 55–64

Kato T, Dong B, Ishii K, Kinemuchi H (1986) Brain dialysis: in vivo metabolism of dopamine and serotonin by monoamine oxidase A but not B in the striatum of unrestrained rats. J Neurochem 46: 1277–1282

Knoll J (1986) The pharmacology of (−)deprenyl. J Neural Transm [Suppl]22: 75–89

Knoll J (1992) (−)Deprenyl medication: a strategy to modulate the age-related decline of the striatal dopaminergic system. J Am Geriatr Soc 40: 839–847

Kuroki T, Tsutsumi T, Hirano M, Matsumoto T, Tatebayashi Y, Nishiyama K, Uchimura H, Shiraishi A, Nakahara T, Nakamura K (1990) Behavioral sensitization to beta-phenylethylamine (PEA): enduring modifications of specific dopaminergic neuron systems in the rat. Psychopharmacol 102: 5–10

Lees A, Kohout L, Shaw JK, Stern G, Elsworth JD, Sandler M, Youdim MBH (1977) Deprenyl in Parkinson's disease. Lancet 2: 791–795

O'Carroll AM, Fowler CJ, Phillips JP, Tobia I, Tipton KF (1983) The deamination of dopamine by human brain monoamine oxidase: specificity for the two enzyme forms in seven brain regions. Naunyn Schmiedebergs Arch Pharmacol 322: 198–202

Otsaka S, Kobayashi Y (1964) A radioisotopic assay for monoamine oxidase determinations in human plasma. Biochem Pharmacol 13: 995–1006

Paterson IA, Juorio AV, Berry MD, Zhu MY (1991) Inhibition of monoamine oxidase-B by (−)deprenyl potentiates neuronal responses to dopamine agonists but does not inhibit dopamine catabolism in the rat striatum. J Pharmacol Exp Ther 258: 1019–1024

Riederer P, Youdim MBH (1986) Brain monoamine oxidase activity and monoamine metabolism in Parkinsonian patients treated with 1-deprenyl. J Neurochem 46: 1349–1356

Riederer P, Konradi C, Schay V, Kienzel E, Youdim MBH (1986) Location of MAO A and MAO B in human brain. A step in understanding the thepeutic action of 1-deprenyl. Adv Neurol 45: 111–119

Roy E, Bedard PJ (1993) Deprenyl increases survival of rat foetal nigral neurones in culture. Neuroreport 4: 1183–1186

Salo PT, Tatton WG (1992) Deprenyl reduces the death of motoneurons caused by axotomy. J Neurosci Res 31: 394–400

Shimoda K, Sauve Y, Marini A, Schwartz JP, Commissiong JW (1992) A high yield of tyrosine hydroxylase-positive cells from rat E14 mesencephalic cell culture. Brain Res 585: 319–331

Simpson LL (1978) Evidence that deprenyl, a type B monoamine oxidase inhibitor, is an indirectly acting sympathomimetic amine. Biochem Pharmacol 27: 1591–1595

Takeshima T, Johnston JM, Commissiong JW (1994) Oligodendrocyte type 2 astrocyte (O-2A) progenitors increase the survival of rat mesencephalic dopaminergic neurons from death induced by serum dperivation. Neurosci Lett 166: 178–182

Tatton WG (1993) Selegiline can mediate neuronal rescue rather than neuronal protection. Mov Disord 8 [Suppl] 1: S20–S30

Tatton WG, Greenwood CE (1991) Rescue of dying neurons: a new action for deprenyl in MPTP parkinsonism. J Neurosci Res 30: 666–672

Tatton WG, Ju WYL, Holland DP, Tai C, Kwan M (1994) (−)-Deprenyl reduces PC12 cell apoptosis by inducing new protein synthesis. J Neurochem 63: 1572–1575

Tekes K, Tothfalusi L, Gaal J, Magyar K (1988) Effect of MAO inhibitors on the uptake and metabolism of dopamine in rat and human brain. Pol J Pharmacol 40: 653–658

Westlund KN, Denney RM, Kochersberger LM (1985) Distinct monoamine oxidase A and B populations in primate brain. Science 230: 181–183

Authors' address: Prof. J. P. M. Finberg, Pharmacology Unit, Rappaport Faculty of Medicine, Technion, POB 9649, Haifa, 31096, Israel

J Neural Transm (1996) [Suppl] 48: 103–112

Potential of neurotrophic factors in therapy of Parkinson's disease

J. C. Möller, J. Sautter, and **A. Kupsch**

Department of Neurology, University of Munich, Federal Republic of Germany

Summary. Neurotrophic factors of dopaminergic neurons may represent a potential neuroprotective therapy for PD. This article reviews published experiments that demonstrate the effects of neurotrophic factors on dopaminergic neurons in vitro and in vivo. At present this issue is predominantly investigated in basic neuroscientific research. Its possible future clinical relevance is discussed.

Introduction

During development neurons are abundantly generated in vertebrate nervous system. But only a proportion of these neurons survive after axons have reached their target area. This is thought to be the consequence of successful competition of the surviving neurons for a specific target-derived, retrogradely transported neurotrophic factor, present only in limited amounts in the target fields. Nerve growth factor has to be considered the prototype of such a neurotrophic factor. Apart from NGF the family of the so-called "neurotrophins" subsumes at present brain-derived neurotrophic factor (BDNF), neurotrophin-3 (NT-3) and neurotrophin-4/5 (NT-4/5). However, the concept of neurotrophic factors as specific, target-derived molecules, each acting on distinct neuronal types, has to be modified because of their high degree of pleiotropism and a considerable overlap in biological activities (for review see Korsching, 1993). In addition, a number of other nontarget-derived molecules exerts trophic actions on certain neurons and nonneuronal cells. These factors include ciliary neurotrophic factor (CNTF), fibroblast growth factors 1 and 2 (FGF-1/FGF-2), insulin-like growth factors 1 and 2 (IGF-1/IGF-2), muscle-derived differentiation factor (MDF) and members of the transforming growth factor β-superfamily such as glial cell line-derived neurotrophic factor (GDNF). Several observations in experimental and clinical research suggest the existence of a trophic factor or factors for dopaminergic neurons. In line with that trophic actions of some of these factors on dopaminergic neurons have been documented in vitro and in vivo and are subsequently reviewed.

Glial cell line-derived neurotrophic factor (GDNF)

GDNF was purified and cloned in 1993, identified by its promotion of survival and morphological differentiation of dopaminergic neurons and increase of their high-affinity dopamine uptake in embryonic midbrain cultures (Lin et al., 1993). Furthermore, specific trophic actions have been described for embryonic motor neurons in vitro as well as in vivo (Henderson et al., 1994; Oppenheim et al., 1995; Yan et al., 1995). In vivo, it was shown that GDNF acts on transmitter activity in midbrain dopaminergic pathways and on spontaneous and amphetamine-induced motor behavior in unlesioned rats and induces sprouting of tyrosine hydroxylase-immunoreactive (TH-IR) neurites towards its injection site suggesting a role for GDNF as a neurotrophic factor for adult dopaminergic neurons in rats (Hudson et al., 1995). In the 6-hydroxydopamine (6-OHDA) lesion model of the rat neurochemical and behavioral improvements following intranigral administration of GDNF were demonstrated (Hoffer et al., 1994). Following hemitransection of the medial forebrain bundle in rat a significant protection of TH-IR nigral neurons by GDNF was reported (Beck et al., 1995a). Daily injections of 5 μg (1 μg) GDNF for 14 days above the substantia nigra led to a survival of 84% (82%) of TH-IR nigral cells, whereas in control animals only 50% of these cells survived. In the 1-methyl-4-phenyl-1,2,3,5-tetrahydropyridine (MPTP)-mouse model of PD the protective effects of GDNF were investigated in respect to the timepoint of its administration (Tomac et al., 1995). Striatal injections of GDNF (10 μg on two consecutive days), 24 hours before MPTP exposure, significantly reduced the normally observed TH-IR cell loss of 30% to about 15%. In addition, this effect was accompanied by a protection of density of dopaminergic nerve terminals and dopamine levels in striatum as well as by increased motor behavior. When GDNF was administered 7 days after MPTP exposure, there was still some recovery of TH-IR cells, increased locomotor behavior and partial restoration of dopamine levels. However, implantation of encapsulated GDNF-producing cells in rats with unilateral dopamine depletion and parkinsonian symptoms does not decrease apomorphine-induced rotations (Lindner et al., 1995). Nonetheless a TH-positive fiber ingrowth into the membranes of the applied capsules was observed. The most encouraging results stem from a recent investigation employing an animal model with a delayed cell death of nigral dopamine neurons (Sauer et al., 1994, 1995). To this end, animals received an unilateral injection of 6-OHDA into the striatum resulting in nigral degeneration with onset at one week and an extensive death of nigral neurons 4 weeks post lesion. Administration of GDNF for 4 weeks over the substantia nigra at a cumulative dose of 140 μg, starting on the day of lesion, completely prevented nigral cell death and atrophy. A partial protective effect was observed as a consequence of one single injection of 10 μg GDNF at one week post-lesion. In conclusion, GDNF is protective in mechanical and toxic lesions and does represent the most promising candidate for growth factor therapy in PD at present. However, possible adverse effects of GDNF, for instance sprouting or aberrant growth in non-diseased systems, remain to be studied.

Brain-derived neurotrophic factor (BDNF)

In vitro, BDNF revealed distinct effects on survival, morphological differentiation, neuritic growth, protection against MPP$^+$ cytotoxicity and dopamine uptake of fetal mesencephalic dopaminergic neurons (Beck et al., 1993; Hyman et al., 1991; Knüsel et al., 1991; Spina et al., 1992; Studer et al., 1995). In a preliminary report this neurotrophin was shown to be retrogradely transported to substantia nigra after injection into the striatum indicating that there are functional receptors for BDNF on adult dopaminergic nigrostriatal neurons (Wiegand et al., 1991). Additionally, BDNF has been reported to prevent cell death of axotomized spinal motor neurons in vivo (Yan et al., 1992). However, in vivo studies on the neuroprotective function of BDNF on dopaminergic neurons are at present contradictory: The reduction of TH-IR nigral cells in rats after hemitransection of the nigrostriatal forebrain bundle was not prevented by intraventricular infusion of BDNF (Knüsel et al., 1992). Furthermore, no improvement of survival of transplanted fetal ventral mesencephalic neurons by daily intrastriatal injection or chronic intraventricular infusion of BDNF in the 6-OHDA-model of the rat was observed (Sauer et al., 1993). Moreover, mice lacking brain-derived neurotrophic factor showed no affection of survival of midbrain dopaminergic neurons (Ernfors et al., 1994). In contrast, BDNF-secreting fibroblasts which were implanted near the substantia nigra 7 days before MPP+-infusion markedly increased nigral dopaminergic neuronal survival in the rat (Frim et al., 1994). Additionally, it has been reported that intranigral BDNF infusions increased amphetamine-induced rotations and enhanced striatal dopamine metabolism suggesting a presynaptic effect of BDNF on nigrostriatal dopamine system (Altar et al., 1992). In summary, experimental evidence concerning the neurotrophic effect of BDNF on nigral dopaminergic neurons is controversial. Thus, at present clinical studies with BDNF in PD patients do not seem to be warranted.

Nerve growth factor (NGF)

In vitro, no trophic effects of NGF on mesencephalic cultures have been shown so far (Knüsel et al., 1990). In vivo, NGF does not protect axotomized nigro-striatal neurons in the adult rat (Knüsel et al., 1992). On the other hand intraventricular injections of NGF were reported to increase striatal dopamine contents in MPTP-treated mice (Garcia et al., 1992). NGF might be of interest because of its trophic actions on sympathetic ganglia cells. In vitro, catecholaminergic adrenal medullary cells resemble sympathic ganglia cells after application of NGF (Anderson et al., 1986). An improved survival of transplanted adult adrenal medulla cells after simultaneous, intraventricular application of NGF in 6-OHDA lesioned rats was observed (Strömberg et al., 1985). In non-human primates a trophic support for grafted rhesus adrenal chromaffin cells by cografting excised peripheral nerve as a source for NGF delivery was shown (Kordower et al., 1990). Clinically, after implantation of autologous adrenal medulla one patient received intraputaminal infusions of

NGF followed by a moderate improvement of clinical symptoms 7 months after surgery (Olson et al., 1991). No follow-up has been published. Corresponding studies have been initiated in the USA, but with discouraging results (Goetz C, personal communication).

Neurotrophin-3 and Neurotrophin-4/5 (NT-3 and NT-4/5)

Apart from BDNF and NGF possible trophic actions of other neurotrophins on dopaminergic neurons remain to be investigated. However, recently NT-3 and NT-4/5 were shown to influence morphological differentiation of rat mesencephalic dopaminergic neurons in vitro (Studer et al., 1995). NT-3 mRNA was detected in substantia nigra by in situ hybridization (Gall et al., 1992). Contradictory results exist in respect to whether or not NT-3 promotes survival of mesencephalic dopaminergic neurons in vitro (Hyman et al., 1994; Knüsel et al., 1991). NT-4/5 elicited an 7-fold increase in the number of cultured dopaminergic neurons as well as an augmentation in dopamine content. In contrast, NT-4/5 had no effect on dopamine uptake capacity (Hyman et al., 1994). Additionally, two-week supranigral infusions of NT-4/5 were shown to elevate the turnover of dopamine through both metabolic and release pools and augment the behavioral response to d-amphetamine in rats (Altar et al., 1994).

Ciliary neurotrophic factor (CNTF)

CNTF exerts protective and neurotrophic functions on embryonic and post-natal lesioned motoneurons in vitro and in vivo. It promotes survival of a wide range of other embryonic neurons in peripheral and central nervous system in vitro (for review see Unsicker et al., 1992). The most striking features of CNTF until 1995 are the in vivo prevention of lesion-induced degeneration of facial motoneurons after axotomy in newborn rats and the antagonization of neurodegenerative changes in the progressive motor neuronopathy (pmn) genetic mouse model of amyotrophic lateral sclerosis (Sendtner et al., 1990, 1992). Furthermore, CNTF was shown to rescue nigral, most likely dopaminergic neurons in the hemitransection model of the rat (as assessed by Nissl-staining), but it did not protect against axotomy-induced reduction of nigral TH-IR cells (Hagg et al., 1993). It would be of considerable interest whether or not the TH-negative neurons will be able to synthezise again this rate-limiting enzyme of dopamine synthesis at a later timepoint. In preclinical trials, the systemic or intrathecal application of CNTF in rats and sheep led to severe side effects including fever, cachexia, disturbance of the blood brain barrier and others (Sendtner, 1995), thereby limiting a possible therapeutical clinical application.

Epidermal growth factor (EGF)

In vitro, EGF was shown to support survival of both embryonic dopaminergic midbrain and cholinergic forebrain neurons (Knüsel et al., 1990). The trophic actions of EGF required the presence of glial cells proposing an indirect mode of action of EGF on dopaminergic neurons. In this context, it was previously reported that EGF acts also on glial cells promoting their ability to proliferate and to differentiate (Leutz et al., 1981; Honegger et al., 1983). In vivo studies revealed that intraventricular administration of EGF in rats 5 weeks after hemitransection of the medial forebrain bundle restores about 20% of TH-IR nigral neurons in comparison to vehicle-treated animals (Pezzoli et al., 1991). Moreover, intraventricular infusion of EGF also accelerates recovery of striatal dopaminergic parameters, i.e. the dopamine content and TH activity, in the MPTP-mouse model (Hadjiconstantinou et al., 1991). However, corresponding non-human primate studies have not been performed so far. Further investigations should also evaluate possible glial reactions induced by application of EGF.

Fibroblast growth factor-2 (FGF-2)

FGF-2 (previous denotion: basic fibroblast growth factor) exhibits trophic effects on central embryonic dopaminergic and GABA-ergic neurons in culture which appear to be glia-mediated (Engele et al., 1991). Transient increases in the amounts of FGF-2 have been described in distinct lesion paradigms of the central nervous system (CNS) (Frautschy et al., 1991; Gomez-Pinilla et al., 1992). The same was suggested to occur in a MPTP-induced lesion of the nigrostriatal dopaminergic system in mice (Leonard et al., 1993). This may indicate a possible role of FGF-2 in neuronal regeneration, for instance in form of induction of synthesis of NGF or of other trophic molecules in astrocytes (Yoshida et al., 1991). In 6-OHDA-lesioned rats intrastriatal FGF-2 infusions neither prevent striatal dopamine depletions nor diminish behavioral deficits (Otto et al., 1992). In the MPTP-mouse model intrastriatal application of FGF-2 via gel foam partially attenuated the toxin-induced damage (Otto et al., 1990). However, this effect was only observed if FGF-2 was applied simultaneously or 3 days after intraperitoneal MPTP-injection, whereas a delay of FGF-2 administration for 7 days after MPTP-injection aborted restoration of transmitters and TH- levels. Apart from these findings, FGF-2 ameliorates rotational behavior of substantia nigra-transplanted rats with lesions of the nigrostriatal dopaminergic system (Matsuda et al., 1992). Results of ongoing non-human primate studies remain to be awaited.

Insulin-like growth factor-1 (IGF-1)

In vitro, IGF-1 was shown to stimulate dopamine uptake of fetal dopaminergic mesencephalic neurons (Knüsel et al., 1990). In addition, the predominant form of IGF-1 in the CNS, des-IGF-1, was very effective in promoting survival of cultured mesencephalic neurons (Beck et al., 1993). In vivo, neurotrophic actions of IGF-1 have to our knowledge not been demonstrated. However, IGF-1 gene disruption does not affect the number of mesencephalic dopaminergic neurons (Beck et al., 1995b), and there was no IGF-1 binding in rat substantia nigra or striatum measured by autoradiography (Araujo et al., 1989). These findings are discouraging in respect to a potential therapeutic effect of IGF-1 in treatment of PD.

Muscle-derived differentiation factor (MDF)

In vitro, MDF induces tyrosine hydroxylase-expression in a variety of CNS neurons, including those of striatum, cerebellum and cortex (Iacovitti, 1991). Normally, i.e. without MDF, these neurons do not express this enzyme of catecholamine synthesis. Further in vitro studies revealed that MDF enhances TH-mRNA 40-fold in fetal mesencephalic neurons (Iacovitti et al., 1992). Preliminary in vivo studies employing infusion of partially isolated MDF reported this molecule to enhance TH-activity in dopamine-depleted striata of 6-OHDA-lesioned rats (Jin et al., 1991). Furthermore, an increase of striatal dopamine concentrations and a partial compensation of rotational asymmetry were observed. In contrast, dopaminergic parameters were not affected by administration of MDF in control rats suggesting that adult dopaminergic neurons may regain sensitivity towards differentiation factors after lesion.

Transforming growth factor-β (TGF-β)

GDNF was identified as a member of the TGF-β superfamily. Therefore the potential trophic actions on dopaminergic neurons of other members of this superfamily attracted attention: It has been shown that TGF-β1, TGF-β2, TGF-β3 and activin A exert a survival-promoting activity on cultured dopaminergic neurons of the developing substantia nigra (Poulsen et al., 1994; Krieglstein et al., 1995). In addition, TGF-β2 and TGF-β3 mRNAs were detected in developing rat striatum and substantia nigra (Poulsen et al., 1994). However, TGF-β3 did not prevent delayed degeneration of nigral dopaminergic neurons following intrastriatal 6-OHDA lesion (Sauer et al., 1995).

Conclusion

Neurodegeneration in PD is a slow and progressive process that occurs over years before any symptoms of the disease appear. Most of the so far utilized animal models, however, lead to a rapid onset of degenerative events. Therefore it is of interest that after intrastriatal 6-OHDA lesion a delayed cell death of nigral neurons may be observed. Although the prevention of this delayed nigral cell death by in vivo application of GDNF features a promising attempt for a possible clinical use of neurotrophic factors in the future, adequate data in respect to the effects of for instance GDNF on non-dopaminergic neurons are not available. The mode of administration represents another major problem in "growth factor therapy" of PD, since the growth factor-proteins possess a relatively high molecular weight preventing them to cross the blood-brain-barrier. An already tested strategy of delivery may be the implantation of genetically modified cells which are able to secrete neurotrophic factors (Lindner et al., 1995). Additionally, one prerequisite for treating PD patients with neurotrophic factors appears to be an early diagnosis of the disease. This is in principle possible with ^{18}F-fluorodopa-PET and βCIT-SPECT, although corresponding prospective studies are not available on a larger scale.

In summary, at present neurotrophic growth factors have not yet been sufficiently investigated in respect to a therapeutic application in PD and correspondingly further experiments including non-human primate studies are necessary to study potential beneficial and adverse effects of long-term application of neurotrophic factors before clinical studies should be initiated to a larger extent.

References

Altar CA, Boylan CB, Jackson C, Hershenson S, Miller J, Wiegand SJ, Lindsay RM, Hyman C (1992) Brain-derived neurotrophic factor augments rotational behavior and nigrostriatal dopamine turnover in vivo. Proc Natl Acad Sci USA 89: 11347–11351

Altar CA, Boylan CB, Fritsche M, Jackson C, Hyman C, Lindsay RM (1994) The neurotrophins NT-4/5 and BDNF augment serotonin, dopamine, and GABAergic systems during behaviorally effective infusions to the substantia nigra. Exp Neurol 130: 31–40

Anderson DJ, Axel R (1986) A bipotential neuroendocrine precursor whose choice of cell fate is determined by NGF and gluco-corticoids. Cell 47: 1079–1090

Araujo DM, Lapchak PA, Collier B, Chabot JG, Quirion R (1989) Insulin-like growth factor −1 (somatomedin C) receptors in the rat brain: distribution and interaction with the hippocampal cholinergic system. Brain Res 484: 130–139

Beck KD, Knüsel B, Hefti F (1993) The nature of the trophic action of brain-derived neurotrophic factor, des(1–3)-Insulin-like growth factor−1, and basic fibroblast growth factor on mesencephalic dopaminergic neurons developing in culture. Neuroscience 52: 855–866

Beck KD, Valverde J, Alexi T, Poulsen K, Moffet B, Vandlen RA, Rosenthal A, Hefti F (1995a) Mesencephalic dopaminergic neurons are protected by GDNF from axotomy-induced degeneration in the adult brain. Nature 373: 339–341

Beck KD, Powell-Braxton L, Widmer HR, Valverde J, Hefti F (1995b) IGF-1 gene disruption results in reduced brain size, CNS hypomyelination and loss of hippocampal granule and striatal parvalbumin-containing cells. Neuron 14: 717–730

Engele J, Bohn MC (1991) The neurotrophic effects of fibroblast growth factor in vitro are mediated by mesencephalic microglia. J Neurosci 11: 3070–3078

Ernfors P, Lee KF, Jaenisch R (1994) Mice lacking brain-derived neurotrophic factor develop with sensory deficits. Nature 368: 147–150

Frautschy SA, Walicke PA, Baird A (1991) Localization of basic fibroblast growth factor and its mRNA after CNS injury. Brain Res 553: 291–299

Frim DM, Uhler TA, Galpern WR, Beal MF, Breakefield XO, Isacson O (1994) Implanted fibroblasts genetically engineered to produce brain-derived neurotrophic factor prevent 1-methyl-4-phenylpyridinium toxicity to dopaminergic neurons in the rat. Proc Natl Acad Sci USA 91: 5104–5108

Gall CM, Gold SJ, Isackson PJ, Seroogy KB (1992) Brain-derived neurotrophic factor and neurotrophin-3 mRNAs are expressed in ventral midbrain regions containing dopaminergic regions. Mol Cell Neurosci 3: 56–63

Garcia E, Rios C, Sotelo J (1992) Ventricular injection of nerve growth factor increases dopamine content in the striata of MPTP-treated mice. Neurochem Res 17: 979–982

Gomez-Pinilla F, Lee JWK, Cotman CW (1992) Basic FGF in the adult brain: cellular distribution and response to entorhinal lesion and fimbria-fornix transection. J Neurosci 12: 345–355

Hadjiconstantinou M, Fitkin JG, Dalia A, Neff NH (1991) Epidermal growth factor enhances striatal dopaminergic parameters in the 1-methyl-4-phenyl-1,2,3,6-tetrahydropyridine-treated mouse. J Neurochem 57: 479–482

Hagg T, Varon S (1993) Ciliary neurotrophic factor (CNTF) prevents axotomy-induced degeneration of adult rat substantia nigra dopaminergic neurons. Proc Natl Acad Sci USA 90: 6315–6319

Henderson CE, Phillips HS, Pollock RA, Davies AM, Lemeulle C, Armanini M, Simpson LC, Moffet B, Vandlen RA, Koliatsos VE, Rosenthal A (1994) GDNF: a potent survival factor for motoneurons present in peripheral nerve and muscle. Science 266: 1062–1066

Hoffer BJ, Hoffmann A, Bowenkamp K, Huettl P, Hudson J, Martin D, Lin LFH, Gerhardt GA (1994) Glial cell line-derived neurotrophic factor reverses toxin-induced injury to midbrain dopaminergic neurons in vivo. Neurosci Lett 182: 107–111

Honegger P, Guentert-Lauber B (1983) Epidermal growth factor (EGF) stimulation of cultured brain cells. I. Enhancement of the developmental increase in glial enzymatic activity. Dev Brain Res 11: 245–251

Hudson J, Granholm AC, Gerhardt GA, Henry MA, Hoffmann A, Biddle P, Leela NS, Mackerlova L, Lile JD, Collins F, Hoffer BJ (1995) Glial cell line-derived neurotrophic factor augments midbrain dopaminergic circuits in vivo. Brain Res Bull 36: 425–432

Hyman C, Hofer M, Barde YA, Juhasz M, Yancopoulos GD, Squinto SP, Lindsay RM (1991) BDNF is a neurotrophic factor for dopaminergic neurons of the substantia nigra. Nature 350: 230–232

Hyman C, Juhasz M, Jackson C, Wright P, Ip NY, Lindsay RM (1994) Overlapping and distinct actions of the neurotrophins BDNF, NT-3, and NT-4/5 on cultured dopaminergic and GABAergic neurons of the ventral mesencephalon. J Neurosci 14: 335–347

Iacovitti L (1991) Effects of a novel differentiation factor on the development of catecholamine traits in noncatecholamine neurons from various regions of the rat brain: studies in tissue culture. J Neurosci 11: 2403–2409

Iacovitti L, Evinger MJ, Stull ND (1992) Muscle-derived differentiation factor increases expression of the tyrosine hydroxylase gene and enzyme activity in cultured dopamine neurons from the rat midbrain. Mol Brain Res 16: 215–222

Jin BK, Schneider JS, Du YY, Iacovitti L (1991) MDF, a muscle factor, produces partial motor recovery in 6-hydroxydopamine lesioned rats by increasing tyrosine hydroxylase activity and catechol levels. Soc Neurosci Abstr 18: 1296

Knüsel B, Michel PP, Schwaber JS, Hefti F (1990) Selective and nonselective stimulation of central and cholinergic and dopaminergic development in vitro by nerve growth factor, basic fibroblast growth factor, epidermal growth factor, insulin and the insulin-like growth factors I and II. J Neurosci 10: 558–570

Knüsel B, Winslow JW, Rosenthal A, Burton LE, Seid DP, Nikolics K, Hefti F (1991) Promotion of central cholinergic and dopaminergic neuron differentiation by brain-derived neurotrophic factor but not neurotrophin 3. Proc Natl Acad Sci USA 88: 961–965

Knüsel B, Beck KD, Winslow JW, Rosenthal A, Burton LE, Widmer HR, Nikolics K, Hefti F (1992) Brain-derived neurotrophic factor administration protects basal forebrain cholinergic but not nigral dopaminergic neurons from degenerative changes after axotomy in the adult rat brain. J Neurosci 12: 4391–4402

Kordower JH, Fiandaca MS, Notter MFD, Hansen JT, Gash DM (1990) NGF-like trophic support from peripheral nerve for grafted rhesus adrenal chromaffin cells. J Neurosurg 73: 413–428

Korsching S (1993) The neurotrophic factor concept: a reexamination. J Neurosci 13: 2739–2748

Krieglstein K, Suter-Crazzolara C, Fischer WH, Unsicker K (1995) TGF-beta superfamily members promote survival of midbrain dopaminergic neurons and protect them against toxicity. EMBO J 14: 736–742

Leonard S, Luthman D, Logel J, Luthman J, Antle C, Freedman R, Hoffer B (1993) Acidic and basic fibroblast growth factor mRNAs are increased in striatum following MPTP-induced dopamine neurofiber lesion: assay by quantitative PCR. Mol Brain Res 18: 275–284

Leutz A, Schachner M (1981) Epidermal growth factor stimulates DNA-synthesis of astrocytes in primary cerebellar cultures. Cell Tissue Res 220: 393–404

Lin LFH, Doherty J, Lile J, Bektesh S, Collins F (1993) GDNF: a glial cell line-derived neurotrophic factor for midbrain dopaminergic neurons. Science 260: 1130–1132

Lindner MD, Winn SR, Baetge EE, Hammang JP, Gentile FT, Doherty E, McDermott PE, Frydel B, Ullman MD, Schallert T, Emerich DF (1995) Implantation of encapsulated catecholamine and GDNF-producing cells in rats with unilateral dopamine depletions and parkinsonian symptoms. Exp Neurol 132: 62–76

Matsuda S, Saito H, Nishiyama N (1992) Basic fibroblast growth factor ameliorates rotational behavior of substantia nigra-transplanted rats with lesions of the dopaminergic nigrostriatal neurons. Jpn J Pharmacol 59: 365–370

Olson L, Backlund EO, Ebendal T, Freedman R, Hamberger B, Hansson P, Hoffer B, Lindblom U, Meyerson B, Strömberg I, Sydow O, Seiger A (1991) Intraputaminal infusion of nerve growth factor to support adrenal medullary autografts in Parkinson's disease. Arch Neurol 48: 373–381

Oppenheim RW, Houenou LJ, Johnson JE, Lin LFH, Li L, Lo AC, Newsome AL, Prevette DM, Wang S (1995) Developing motoneurons rescued from programmed and axotomy-induced cell death by GDNF. Nature 373: 344–346

Otto D, Unsicker K (1990) Basic FGF reverses chemical and morphological deficits in the nigrostriatal system of MPTP-treated mice. J Neurosci 10: 1912–1921

Otto D, Unsicker K (1992) Effects of FGF-2 on dopaminergic neurons. Neurosci Facts 3: 82–83

Pezzoli G, Zecchinelli A, Ricciardi S, Burke RE, Fahn S, Scarlato G, Carenzi A (1991) Intraventricular infusion of epidermal growth factor restores dopaminergic pathways in hemiparkinsonian rats. Mov Disord 6: 281–287

Poulsen KT, Armanini MP, Klein RD, Hynes MA, Phillips HS, Rosenthal A (1994) TGF beta 2 and TGF beta 3 are potent survival factors for midbrain dopaminergic neurons. Neuron 13: 1245–1252

Sauer H, Oertel WH (1994) Progressive degeneration of nigrostriatal dopamine neurons following intrastriatal terminal lesions with 6-hydroxydopamine: a combined retrograde tracing and immunocytochemical study in the rat. Neuroscience 59: 401–415

Sauer H, Fischer W, Nikkah G, Wiegand P, Brundin P, Lindsay RM, Björklund A (1993) Brain-derived neurotrophic factor enhances function rather than survival of intrastriatal ventral mesencephalic grafts. Brain Res 626: 37–44

Sauer H, Rosenblad C, Björklund A (1995) GDNF but not TGFβ3 prevents delayed degeneration of nigral dopaminergic neurons following striatal 6-hydroxydopamine-lesion. Proc Natl Acad Sci USA 92: 8935–8939

Sendtner M (1995) Neurotrophic factors for motoneurons. J Neurol 242, S2:S1

Sendtner M, Kreutzberg GW, Thoenen H (1990) Ciliary neurotrophic factor prevents the degeneration of motor neurons after axotomy. Nature 345: 440–441

Sendtner M, Schmalbruch H, Stöckli KA, Caroll P, Kreutzberg GW, Thoenen H (1992) Ciliary neurotrophic factor prevents degeneration of motor neurons in mouse mutant progressive motor neuronopathy. Nature 358: 502–504

Spina MB, Squinto SP, Miller J, Lindsay RM, Hyman C (1992) Brain-derived neurotrophic factor protects dopamine neurons against 6-hydroxydopamine and N-methyl-4-phenylpyridinium ion toxicity: involvement of the glutathione system. J Neurochem 59: 99–106

Strömberg I, Herrera-Marschitz M, Ungerstedt U, Ebendal T, Olson L (1985) Chronic implants of chromaffin tissue into the dopamine-denervated striatum. Effects on NGF on graft survival, fiber growth and rotational behaviour. Exp Brain Res 60: 335–349

Studer L, Spenger C, Seiler RW, Altar A, Lindsay RM, Hyman C (1995) Comparison of the effects of the neurotrophins on the morphological structure of dopaminergic neurons in cultures of rat substantia nigra. Eur J Neurosci 7: 223–233

Tomac A, Lindquist E, Lin LFH, Ögren SO, Young D, Hoffer BJ, Olson L (1995) Protection and repair of the nigrostriatal dopaminergic system by GDNF in vivo. Nature 373: 335–339

Unsicker K, Grothe C, Westermann R, Wewetzer K (1992) Cytokines in neural regeneration. Curr Opin Neurobiol 2: 671–678

Wiegand SJ, Alexander C, Lindsay RM, DiStefano PS (1991) Soc Neurosci Abstr 17: 1121

Yan Q, Elliot J, Snider WD (1992) Brain-derived neurotrophic factor rescues spinal motor neurons from axotomy-induced cell daeth. Nature 360: 753–755

Yan Q, Matheson C, Lopez QT (1995) In vivo neurotrophic effects of GDNF on neonatal and adult facial motor neurons. Nature 373: 341–344

Yoshida K, Gage F (1991) Fibroblast growth factors stimulate nerve growth factor synthesis and secretion by astrocytes. Brain Res 538: 118–126

Authors' address: Dr. J. C. Möller, Department of Neurology, University of Munich, Marchioninistrasse 15, D-81377 Munich, Federal Republic of Germany

SpringerNeurology

K. A. Jellinger, M. Windisch (eds.)

New Trends in the Diagnosis and Therapy of Non-Alzheimer's Dementia

1996. 61 partly coloured figures. VIII, 288 pages. Soft cover DM 190,–, öS 1330,–
Reduced price for subscribers to "Journal of Neural Transmission":
Soft cover DM 171,–, öS 1197,–. ISBN 3-211-82823-0
Journal of Neural Transmission, Supplement 47

This volume gives an overview of the present state of art on the classification, neuropathology, clinical presentation, neuropsychology, diagnosis, neuroimaging and therapeutic possibilities in non-Alzheimer's dementias, an increasingly important group of CNS diseases, which account for 7 to 30% of dementing disorders in adults and aged subjects, and thus, represent the second most frequent cause of dementia after Alzheimer's disease. The monograph provides the newest information for neurologists, psychiatrists, dementia research workers, dementia clinicians, neuropathologists, neurobiologists, and practicing physicians.

P. Riederer, W. Wesemann (eds.)

Parkinson's Disease: Experimental Models and Therapy

1995. 121 figures. XI, 466 pages. Soft cover DM 240,–, öS 1680,–
Reduced price for subscribers to "Journal of Neural Transmission":
Soft cover DM 216,–, öS 1512,–. ISBN 3-211-82749-8
Journal of Neural Transmission, Supplement 46

Current research on Parkinson's disease is aimed at the goal of determining the underlying cause of this terrible disease and of developing adequate treatment strategies to deal with it. This volume focuses on models that mirror the progression of the symptoms of Parkinson's disease (iron, MPTP, 6-hydroxydopamine, "TaClo", etc.) while other topics are the evaluation of oxidative stress, calcium, excitotoxicity, nitric oxide, or nerve growth factors as possible pathophysiological candidates or causal parameters. Further topics are the interplay between exogenous and endogenous toxins, the potential of brain imaging by PET, MRI and SPECT, as well as promising therapeutic drug strategies. This volume represents a comprehensive survey of the state of the art for neurologists, biochemists, neuropharmacologists and toxicologists.

SpringerWienNewYork

P.O.Box 89, A-1201 Wien • New York, NY 10010, 175 Fifth Avenue
Heidelberger Platz 3, D-14197 Berlin • Tokyo 113, 3-13, Hongo 3-chome, Bunkyo-ku

SpringerNeurology

U. Bonuccelli, J. M. Rabey (eds.)

Old and New Dopamine Agonists in Parkinson's Disease

1995. 73 figures. VIII, 321 pages. Soft cover DM 180,–, öS 1260,–
Reduced price for subscribers to "Journal of Neural Transmission":
Soft cover DM 162,–, öS 1134,–. ISBN 3-211-82717-X
Journal of Neural Transmission, Supplement 45

This book provides a comprehensive overview of the basic and clinical neuropharmacology of dopamine agonists and the rationale for their employment in PD. The authors have compiled an up-to-date guide, covering such topics as the pathophysiology of dopaminergic systems and the neuro-biochemistry of dopaminergic receptors, the clinical use of old and new dopamine agonists, both in the first-time treatment of PD patients and for reducing motor fluctuations in levodopa-treated ones, and the possible role of dopamine agonists as neuroprotective agents. Particular emphasis has been placed on apomorphine, an old dopamine agonist that has recently recaptured neurologists' interest for its use in both diagnostic use and therapeutic management of advanced parkinsonian patients. Articles discussing the results of ongoing clinical studies of newly developed dopamine agonists and the potential use of dopamine agonists, both new and old, as neuroprotectors should be of particular interest to the reader. The work is an exhaustive up-to-date compendium that assembles the entire spectrum of current basic and clinical research on dopaminergic systems and dopamine agonists in Parkinson's disease into a single authoritative source.

S. Hoyer, D. Müller, K. Plaschke (eds.)

Cell and Animal Models in Aging and Dementia Research

1994. 63 figures. VIII, 272 pages. Soft cover DM 170,–, öS 1190,–
Reduced price for subscribers to "Journal of Neural Transmission":
Soft cover DM 153,–, öS 1071,–. ISBN 3-211-82549-5
Journal of Neural Transmission, Supplementum 44

Although age has been recognized as a risk factor for late-onset dementia of Alzheimer type, its etiology is unknown as yet. Several age-related metabolic abnormalities may thus become important for the pathogenesis of the late-onset form. Studies at the cellular/molecular level in brain tissue are possible post mortem, but lack information on the beginning of the disorder. In this supplement, different approaches are dealt with how to induce structural and/or metabolic abnormalities in relevant cell cultures, in brain slices and in experimental animals, and how behavioral changes parallel the metabolic variations.

SpringerWienNewYork

P.O.Box 89, A-1201 Wien • New York, NY 10010, 175 Fifth Avenue
Heidelberger Platz 3, D-14197 Berlin • Tokyo 113, 3-13, Hongo 3-chome, Bunkyo-ku

Springer-Verlag
and the Environment

WE AT SPRINGER-VERLAG FIRMLY BELIEVE THAT AN international science publisher has a special obligation to the environment, and our corporate policies consistently reflect this conviction.

WE ALSO EXPECT OUR BUSINESS PARTNERS – PRINTERS, paper mills, packaging manufacturers, etc. – to commit themselves to using environmentally friendly materials and production processes.

THE PAPER IN THIS BOOK IS MADE FROM NO-CHLORINE pulp and is acid free, in conformance with international standards for paper permanency.